Studies in

GW01458918

Volume 3

Logic is not Mathematical

Studies in Logic Series Editor
Dov Gabbay

dov.gabbay@kcl.ac.uk

Logic is not Mathematical

Hartley Slater

ISBN 978-1-84890-051-6

College Publications
Scientific Director: Dov Gabbay
Managing Director: Jane Spurr
Department of Informatics
King's College London, Strand, London WC2R 2LS, UK

http://www.collegepublications.co.uk

Original cover design by orchid creative   www.orchidcreative.co.uk
Printed by Lightning Source, Milton Keynes, UK

CONTENTS

CHAPTER 1

INTRODUCTION

1

I have collected the papers in this volume together so that their overall unity and purpose can be better understood. All except two have been published individually in notable places. But their overall strength and force perhaps cannot be appreciated until they are seen and read together.

Together they correct some central aspects of the development of logic since the end of the nineteenth century. For there have been a number of large errors in the subject, not commonly noticed, that have taken it well off track, in places. In fact the errors in question have the remarkable quality that they are so stupendous it is barely credible that so many intelligent people would make them. This is a fact that presumably itself obscures these errors from common view, and even inhibits investigation into them by more probing minds, since it is only recently that they have surfaced. Certainly there have been immense advances in the subject in the time in question that have to be granted. But the subject also went decidedly off the rails in several hitherto unthought-of places, which are detailed in the chapters that follow.

I have written before about how and why this derailment took place — see, for instance, 'Frege's Hidden Assumption' and 'Logic and Grammar'. Both of these papers, which go into the matter more fully, are here reprinted as appendices. It is not the inevitable herd mentality of the generality of practitioners in the textbook tradition, capable of little beyond reproduction and quotation. Nor is it the irresponsible mental orientation of fringe members of the tradition pre-occupied with the pure mathematics of deviant formal systems: they can be ignored, even if they will continue to play their arbitrary games with symbols. The problem has principally been inattention to natural grammar in certain areas, by the central group of influential people that initiated and led the textbook tradition. It is this that has brought the present situation to pass: the people leading the masses having talents that lie, in broad terms, in numeracy rather than literacy. What change can be expected is not at all clear. But since most of the following pieces (and a good many more before) have all been recognised as publishable in respectable and responsible quarters, there is evidently enough support in the wider population for at least a public defence of the present outlook.

In the second appendix, 'Frege's Hidden Assumption', I locate a specific passage in Frege's writings where he made the crucial mistake that led him to Russell's Paradox. Whether by accident or not, it concerns a case that many others have been confused by, in related but different ways: the case of a suicide, 'Cato killed himself'. My point here is that the predicate 'killed himself' contains a pronoun and therefore has a variable sense, i.e. expresses a variable property, and so should not be formalised by a constant predicate. But 'is a suicide', people often remark, although it contains a trace of the pronoun 'self' in 'sui' is an entirely constant property. The point to make, though, is that the property of being a suicide is a property of a different subject than Cato: it is a property of Cato's killing himself, which is an event, or state of affairs. The inability of the mainline logical tradition to construct the nominal phrase 'Cato's killing himself' from the sentence 'Cato killed himself', of course, contributes to the difficulty of locating the exact subject from the current formal perspective. And that has further ramifications, as is shown in many of the chapters in the body of the book. The third appendix, 'Logic and Grammar', covers this matter, as well as the issue of variable properties, but also discusses several other areas where current formal logic's attention to basic grammar has been lacking. The areas covered go beyond the specific topics in this book, and include theories of numbers, theories of sets, and theories of propositions, for instance. But this final appendix ends with an extended general discussion of the reasons why natural grammar has had so little attention by the formal tradition. Amongst other things I conclude: 'Intelligence is clearly not linked to linguistic competence; and the downside of that is that very brainy people can easily have blind spots in areas like grammar'.

The chapters in the body of the book start, in the second chapter, 'Back to Aristotle!', with some points that have been widely missed even about the Aristotelian origins of logic. Some of these points relate to mistakes made in medieval times, but there are even more severe and pervasive mistakes that have appeared during the twentieth century, in the work of Russell, and Quine, for instance, who were elaborating the novelties brought into the subject by Frege — and Peirce. The main insight needed to correct these modern errors is to see the need for a language including Hilbert's epsilon terms, as found in his Epsilon Calculus. These terms provide reference to the individuals in Frege's Predicate Calculus — they are what Russell called 'logically proper names' — and their use illuminates several issues to do with Realism, including how necessary beings are accessed, for instance. Amongst other things, that is a point that is turned, in the last chapter in the body of the book, chapter eleven, 'Logic is not Mathematical', to help show just why logic is a literary subject, and not a mathematical one.

The third chapter, 'Completing Russell's Logic', details the specific formal elements that distinguish Hilbert's calculus from the Fregean one that is more commonly studied. The formal technology underlying the metaphysical points about Realism involves providing descriptive replacements for certain anaphoric pronouns. Standard predicate logic cannot handle such pronouns, because it is not subtle enough. Several cases are considered including some 'donkey sentences' that throw a considerable light on the possibility of finding conditionals whose probability is a conditional probability, thus demonstrating Adams' and Stalnaker's Thesis. In addition, the distinction is clarified between properly referential terms and what Russell called 'disguised descriptions', i.e. ordinary proper names, enabling a proof that referential terms in intensional constructions quite generally are transparent, and so not opaque as the Fregean tradition has thought. The distinction is important in other ways, as well, since it enables one to clearly demarcate proper identities from 'contingent identities'. Further details of the differences between the two calculi — both formal and metaphysical — are set out in the fourth chapter, 'Ontological Discriminations', which, amongst other things, extends the distinction between referential and descriptive terms to elements in other categories. It also leads to the resolution of a problem in Modal Logic that has defied previous solution.

That rounds out most of the close attention given to the first theme in the book — although significant use of epsilon terms continues. In particular it continues in the first appendix, 'The Central Error in the Tractatus', which consists in a more specialist study of Wittgenstein's early work, *Tractatus Logico-Philosophicus*, showing how ignorance of the Epsilon Calculus led to the most serious error there, in the theory of identity. A discussion of some of Wittgenstein's later work is to be found in chapter seven, and there again the Epsilon Calculus figures significantly. But by this time, it seems, Wittgenstein had learned informally the basic principle of the epsilon calculus, as is evidenced in his understanding of paradigms. Some quotations from Wittgenstein on other matters also occur at other places in the book, showing not only the extent of his relevance to the topics raised here, but also my personal indebtedness to him. The main remark of his in the present connection needs reiterating at the start, though: 'Mathematical Logic' has completely deformed the thinking of mathematicians and of philosophers, by setting up a superficial interpretation of the forms of our everyday language as an analysis of the structures of facts. It is this I principally hope to substantiate in what follows.

The second theme in the book extends through most of the remaining chapters, and concerns another large omission in the language of mainline twentieth century formal logic: the nominalising functors, mentioned before, that turn sentences into noun phrases referring to states of affairs, or propositions. I show, in particular, for a start, in the fifth chapter, 'Out of the Liar Tangle', how standard difficulties with some central propositional paradoxes are overcome within a language that is extended to incorporate such nominal terms. The paper arose through a consideration of one by Stephen Read on Bradwardine, whose treatment is in some respects quite similar. But it does not use propositional referential terms, and in the sixth chapter, 'Translatable Self-Reference', I go on to provide a more detailed treatment of the same puzzles, showing there is in fact an intractable difficulty with Bradwardine's solution which is readily overcome using natural language's forms of expression. I then go on to provide a broader account, in the next chapter again, chapter seven, 'What Priest has been missing', of how the shift from *talking about* mentioned sentences to *using* the associated noun clauses does the required trick. The extent of the change effected through the introduction of referring terms to propositions includes a re-evaluation of Gödel's Theorems, and the place of formal derivations in proofs of facts. Both issues relate closely to David Hilbert's views about Meta-mathematics, and the Formalist Philosophy they have engendered, and it is here that some of Wittgenstein's later remarks find a place in opposition to Hilbert.

The application of this extended formal language to other problems then becomes the starting point for what follows, and in the eighth and ninth chapters, 'Natural Language Consistency' and 'A Perfect Language?', a variety of other long-standing problems in propositional logic are discussed. The eighth chapter, for instance, sets out the extent of contextuality in natural language, and shows that attempts to escape from it, in the formal languages of recent logic, not only have contributed to insurmountable problems like the Liar Paradox, but also have been a failure in their own right because 'eternal sentences' cannot do the job that was intended for them. The ninth chapter details more fully several new forms of natural language expression, and natural language inference, that become available for formal treatment given the introduction of propositional referential terms. But this chapter ends by setting out how parallel processes and points bear on problems with properties rather than propositions, and so on predicates rather than sentences. The following chapter, chapter ten, 'Quine's other way out', then amplifies the resultant treatment of some standard predicate paradoxes by developing new axioms for Set Theory. The basis for this development might be called a 'new' account of predicates, but it is only 'new' to a limited audience — the tradition that got into the standard paradoxes in Set Theory by ignoring natural grammar in this area. The account of predicates that is developed is in fact as old as natural language, and has only been lost to consciousness amongst those who have not given appropriate attention to their mother tongue, through not being fully literate and articulate.

The body of the book then ends with a summary chapter, chapter eleven, 'Logic is not Mathematical', recapitulating the overall line that the argument has taken throughout, but also extending it to two further areas: Modal Logic and General Intensional Logic. That leads naturally to a theory of fictions, which is contrasted with Graham Priest's recent one. In sum this chapter shows that the errors that have been corrected arose not only through a contempt for ordinary language, but also through the related, more basic failure to recognise that being logical is not a matter of being brainy, but of being coherent. It is not a mathematical talent, but a literary one, this being the overall theme of the book. In particular it is the same conclusion I go on to demonstrate with respect to Intensional Logic at this place, and specifically with respect to literary fictions, since Literature is one of the central items that has been misunderstood through the failure to miss the points made about Aristotle with which the book began.

Bibliographic details of the already published chapters are as follows (some of the original papers have been edited here, to avoid overmuch repetition): 'Back to Aristotle!' *Logic and Logical Philosophy* 2011; 'Completing Russell's Logic' *Russell* 27, 2007, 78-92, reprinted in N. Griffin, D. Jacquette and K. Blackwell (eds) *After 'On Denoting'* (The Bertrand Russell Research Centre, Hamilton, 2007, 144-158) McMaster University; 'Ontological Discriminations' *LOGICA Yearbook 2009* (Czech Institute of Philosophy, Prague, 2010); 'Out of the Liar Tangle' in E. Genot, S. Rahman and T.Tulenheimo (eds) *Unity, Truth and the Liar* (Springer, Berlin, 2008, 187-197) Springer Netherlands; 'Translatable Self-Reference' *Australasian Journal of Logic* 2011 (http://www.philosophy.unimelb.edu.au/ajl/); 'What Priest (amongst many others) has been missing' *Ratio* XXIII, 2010, 184-198 Blackwell Publishing Ltd.; 'Natural Language Consistency' *Logique et Analyse* 2011; 'Quine's other way out' *LOGICA Yearbook 2010* (Czech Institute of Philosophy, Prague, 2011); 'The Central Error in the Tractatus' *Wittgenstein Jahrbuch 2003/2006* (Peter Lang, Bern, 2007, 57-66) Peter Lang; 'Frege's Hidden Assumption' *Critica* 38, 2006, 27-37; 'Logic and Grammar', *Ratio* XX, 2007, 206-218.

CHAPTER 2

BACK TO ARISTOTLE

1

In this chapter I investigate how it came to be that Aristotelian Syllogistic lost ground to the logic developed by Peirce and Frege. This is puzzling because there are very close similarities between Aristotle's original work on singular statements and a main theme in Modern Logic: Russell's Theory of Descriptions. But it also highly curious because the points that have to be made to defend Aristotle are in the main very elementary, and quite well known.

There were already confusions in the Middle Ages with the reading of Aristotle on negative terms, and removing these confusions shows that the four traditional Syllogistic forms of statement can be readily generalised not only to handle polyadic relations (for long a source of difficulty), but even other, more measured quantifiers than just 'all', 'some', and 'no'. But these historic confusions merely supplement the main confusions, which arose in more modern times, regarding the logic of singular statements. These main confusions originate in the inability of the mainline modern tradition to supply the 'logically proper names' which alone have the right to replace individual variables; an inability which has resulted in the widespread, but erroneous replacement of individual variables with ordinary proper names, i.e. names for contingent beings, in many if not most contemporary logic texts. The chapter includes the exhibition and grammatical characterisation of the logically proper names that are required instead, specifying just how they differ syntactically from ordinary proper names. It also shows how ontologically significant is the distinction, since not only do logically proper names refer to necessarily existent objects (showing there are no 'empty domains' for Classical Logic to fail to apply to), but also thereby central features of Realism become considerably clarified.

2

The following was most probably Aristotle's account of the Square of Opposition (see Thompson 1953, although see Prior 1962, 169 for some doubts):

(A) All S is P: $(x)(Sx \supset Px)$ & $(\exists x)Sx$

(i.e. 'Any S is P' + 'there are Ss')

(E) No S is P: $(x)(Sx \supset \neg Px)$

(I) Some S is P: $(\exists x)(Sx \& Px)$

(0) Not all S is P: $(\exists x)(Sx \& \neg Px)$ v $\neg(\exists x)Sx$

(Some S is not P: $(\exists x)(Sx \& \neg Px)$)

As can be seen, it involves separating internal negatives from external ones, but also it can be generalised to more measured quantifiers and to relational expressions, like 'Some boys love every girl', as we shall shortly see.

The Boolean tradition has forgotten Aristotle (and natural language), with the result that the internal/external negation distinction has been lost, and the general forms get misunderstood. Thus 'all' becomes 'any', and 'not all' becomes 'some not'. But also thereby the above generalisations disappear as well. For the above Aristotelian forms are representable in probabilistic terms — (A) $\mathrm{pr}(Px/Sx) = 1$, (I) $\mathrm{pr}(Sx \& Px) > 0$, (E) $\mathrm{pr}(Sx \& Px) = 0$, (O) $\mathrm{pr}(Px/Sx) \neq 1$ — and that allows in other probabilistic expressions.

On this analysis the positive forms A and I carry existential import but the negative forms E and O do not. So the law of Obversion does not hold. Thus XEY does not imply XA¬Y, and XOY does not imply XI¬Y. The lack of implication in the latter case, for instance, is because the O form is now read with an external negation: 'Not all Xs are Ys', in place of 'Some Xs are not Ys', which has an internal or predicate negation. 'Not all Xs are Ys' is then the disjunction of 'Some Xs are not Ys' and 'No Xs are Ys' (or 'There are no Xs'). Also Contraposition does not hold: XAY does not imply ¬YA¬X, and XOY does not imply ¬YO¬X. The former result resolves, amongst other things, Hempel's Paradox of Confirmation, through the non-equivalence between 'All ravens are black' and 'All non-black things are non-ravens'.

All this derives from the fact that Aristotle distinguished external from internal negations. What has come down to us from antiquity is a mix of this account together with additions, principally about negative terms, first provided by Boethius. For Boethius equated external negations with internal ones. This was copied by Buridan, amongst others in the later middle ages, although how the discrepancy with Aristotle survived the re-translation of the original texts initiated by St. Thomas is something of a mystery. But the totality of this tradition is not viable at all, once one takes a more rigorous look at it than the medievals evidently did. Boole, for instance, who initiated the now common 'positive and negative existential' tradition had to abandon most of the relations in the traditional Square of Opposition. More ironically, since Boole was attempting a joint theory of logic and probability, he lost sight of the proper connection between logic and probability in the process.

For there are other, quite decisive reasons why the above interpretation has to be kept. First there is supporting evidence for the interpretation of universal statements when one considers other quantifiers. Thus 'Almost all Xs are Ys', 'Most Xs are Ys', and 'A lot of Xs are Ys' surely all entail 'Some Xs are Ys'. Also 'Not a lot of Xs are Ys', for instance, unlike 'A few Xs are not Ys', allows it to be possible that no Xs are Ys (or that there are no Xs at all). So, unlike when there is an internal negation, there is no entailment from the form with the external negation to 'Some Xs are not Ys'. But clearly, also, a probabilistic analysis of the Aristotelian forms supports the reading. For $\mathrm{pr}(Yx/Xx) = 1$ entails $\mathrm{pr}(Yx \& Xx) > 0$. But $\mathrm{pr}(Yx/Xx) \neq 1$ does not entail $\mathrm{pr}(\neg Yx \& Xx) > 0$, since it is possible that $\mathrm{pr}(Xx) = 0$, in which case the conditional probability is not defined. The probabilistic analysis is applicable also, of course, to many other quantifiers. Thus 'Most Xs are Ys' can be represented as '$\mathrm{pr}(Yx/Xx) >$

1/2', and 'A few Xs are not Ys' can be represented as 'pr($\neg$Yx/Xx) $<$ 1/2'.

But neither other quantifiers, nor a probabilistic interpretation are widely considered. Concentration on existential quantifiers and their negations has taken over most of the public's attention. As a result, one supposed difficulty with Aristotelian Syllogistic has continued to be thought insuperable: its seeming restriction to monadic predicate logic. In this respect the polyadic logics developed by Peirce and Frege were taken to win out. How could Aristotelian Syllogistic be extended to handle relational expressions like 'All boys love some girls', for example? That has been taken to be a great stumbling block. But not only can probabilistic analyses easily handle such cases; they also can be easily generalised to many other quantifiers. Thus 'All boys love some girls' is 'pr([pr(Lxy/Gy) $>$ 0]/Bx) $=$ 1', and 'Most boys love few girls' is 'pr([pr(Lxy/Gy) $<$ 1/2]/Bx) $>$ 1/2'. (If probabilities are measured as proportions of existent cases as in 'balls in urns' examples there is no theoretical problem with the nested probabilities.)

3

But there are even more misunderstandings when we come to look at the singular case. The singular case is where $(x)(y)((Sx \,\&\, Sy) \supset x=y)$:

Socrates is well: $(\exists x)(Sx \,\&\, Wx)$

(also: $(x)(Sx \supset Wx) \,\&\, (\exists x)Sx)$

Socrates is ill: $(\exists x)(Sx \,\&\, Cx \,\&\, \neg Wx)$

(where $(x)(Wx \supset Cx))$

Socrates is not ill: $(x)((Sx \,\&\, Cx) \supset Wx)$

Socrates is not well: $(x)(Sx \supset \neg Wx)$

(also: $(\exists x)(Sx \,\&\, \neg Wx) \vee \neg(\exists x)Sx)$

This was definitely Aristotle's account, involving two contraries spanning a category ('Cx' is perhaps 'x is animate' in the case above). It again involves separating internal negatives from external ones. In general form this was also Russell's account, and Quine's account, as we shall see, and, most importantly, in all cases 'Socrates is well' is not of the form 'Wx'.

The common contemporary analysis of 'Socrates is well' as being of the form 'Wx' forgets Russell and Quine, although it was encouraged by them through their failure to present sentences of the form 'Wx'. Just what is of the form 'Wx' will appear later in this chapter. In brief, a sentence is of the form 'Wx' only if 'x' refers to a necessary existent, which happens if 'x' is an Hilbertian epsilon term (see, for example, Slater 2009b), rather than a Russellian iota term. The distinction has several large consequences.

How do Russell and Quine come into the picture? Aristotle, as above, applied the distinction between external and internal negations to singular statements involving contrary terms, saying specifically that both 'Socrates is well' and 'Socrates is ill' would be false if 'Socrates does not exist' was true. Natural language supports Aristotle: if someone says 'My pen is in the drawer' and you look in the drawer and find no pen, or only the pens of others, then you may say

'It is not in the drawer'; in doing so you leave the existence of the pen open, in contrast to when you say the contrary remark 'It is outside the drawer'. Taking 'Socrates' to be a 'disguised description', and so like 'my pen', in the manner of Russell, this shows that Russell's analysis of definite descriptions (which was also Quine's) followed the same line of analysis (though without explicit reference to contrary terms). Thus, for Russell, 'The king of France is bald' entails 'Some king of France is bald', and it has two negatives. For what we may write 'It is not the case that the king of France is bald' does not entail 'The king of France is not bald'. The former expression contains an external negation, and so a 'secondary occurrence' of the subject term. It is true if the king of France does not exist. The latter expression contains an internal, or predicate negation, and so a 'primary occurrence' of its subject term. It is false if the king of France does not exist, and since kings are in the right category, the 'not bald' in this case can be replaced by the contrary term to 'bald', namely 'hirsute'.

However, in addition to the medieval mix-ups with negative terms, another aspect of Russell's theory seems to have obscured the proper Aristotelian inter-pretation. For in symbols 'The king of France is bald' is, of course, on Russell's account:

$$(\exists x)(Kx \ \& \ (y)(Ky \supset y=x) \ \& \ Bx).$$

'The king of France is not bald' is then the same with '$\neg$' before the last 'B', and 'It is not the case that the king of France is bald' is the same with a '$\neg$' before the initial '$(\exists x)$'. Trouble starts because Russell introduced 'iota terms' to symbolise these things another way. Amongst other things this seems to have influenced the tradition of Free Logic in a way we shall look at briefly later. But it also seems to have affected even more substantially the tradition in Classical, i.e. Fregean Logic. For Russell also wrote, making 'the king of France' look like an individual term, 'The king of France is bald' as '$B\iota yKy$', i.e.

$$(\exists x)(x=\iota yKy \ \& \ Bx).$$

In these terms the form with the internal negative, 'The king of France is not bald' becomes

$$(\exists x)(x=\iota yKy \ \& \ \neg Bx).$$

And the form with the external negative, 'It is not the case that the king of France is bald' becomes

$$\neg(\exists x)(x=\iota yKy \ \& \ Bx).$$

The problem is that in the formal symbolism we only have 'Bx' and its contradictory '$\neg$Bx' as elementary forms, so evidently if elementary statements about 'the king of France' have a fourfold logic then 'the king of France' cannot be identical to any 'x' in the given identities. As should be well known, an ordinary name for a contingent individual such as 'Socrates,' is not a 'logically proper name', and only such are proper substitutes for variables. In Russell's terms, the definite description 'the king of France' is not a 'complete symbol' for an individual. To get the logic right with ordinary names one has to employ individuating descriptions such as 'pegasises' in Quine's example, or simply 'is called Pegasus'.

These points have been well rehearsed, but they are not always well remembered. Because of that they suggest a major contemporary reason why, by contrast, the Aristotelian scheme has been lost sight of in recent times. For, despite Russell and Quine, it is very common to find ordinary proper names in the place of the variables in modern logic texts, both classical and free. This makes it seem that they have just a binary logic, whereas they have a fourfold one, as in Aristotle. And *why* are ordinary proper names so often found in the place of variables? Surely the fault must lie with Russell's use of iota terms in identities. For while insisting that only logically proper names should be put in the place where individual variables go, he not only used the confusing iota-term symbolism, making it seem that such terms were complete symbols, but he also provided no examples of complete symbols. Certainly he made some suggestions about them in his lectures on Logical Atomism, but he never provided a formal account. Quine even held that the quantified forms involving 'pegasises' could not have instantiations, so that the burden of reference was carried just by general pronouns such as 'someone'.

The faultiness of these positions is reinforced once one realises that there is no difficulty in finding the appropriate complete individual terms in Classical Logic. For one classical theorem in predicate logic, for instance, is:

$$(\exists x)((\exists y)Ky \supset Kx).$$

So the existence of certain objects is guaranteed, and in the epsilon calculus they are given names: epsilon terms. Thus

$$(\exists y)Ky \supset K\epsilon xKx,$$

is a theorem there, and

$$(\exists y)(y=\epsilon xKx)$$

is also a theorem there. But the Russellian

$$(\exists y)(y=\iota xKx)$$

still isn't. For it is contingent whether anyone is a sole king of France, but what is not contingent is that there is something that is a sole king of France *if anything is a sole king of France*. In the case of names it is contingent whether there is anything called 'Pegasus', but not contingent that there is something called 'Pegasus' if anything is called 'Pegasus' — $\epsilon x(x$ is called 'Pegasus'). If Russell had presented such complete terms, then no-one following him would have had an excuse to put ordinary proper names in place of variables, and the fourfold logic of ordinary proper names, and much more, would have been better appreciated.

4

For many other things are revealed once we have logically proper names in the form of epsilon terms. Thus it is sometimes said that Classical Logic 'assumes the domain is not empty', and sometimes free logics are constructed that supposedly lack this assumption. How can one get, for instance, 'Fa', for some 'a', from '(x)Fx' couldn't there be no entities for 'x' to range over in the universal quantifier? But 'there is a king' is entailed by 'it is a king' (no matter what 'it' is), and also entails 'it is a king' if the 'it' refers to one of the objects alluded to in the existential remark. It is this necessary equivalence that is formalised in the epsilon calculus theorem:

$(\exists x)Kx \equiv K\epsilon xKx$.

That shows there is invariably some object associated with an existential remark, but by introduction of negations that theorem also means that

$(x)Kx \equiv [\neg(\exists x)\neg Kx \equiv \neg\neg K\epsilon x\neg Kx \equiv] K\epsilon x\neg Kx$.

So there is also invariably some object associated with a universal remark, and
'$\epsilon x\neg Kx$' will do as the 'a' needed above. That, of course, still leaves free logics
to be constructed where the related 'a's are *ordinary proper names*, although
most extant free logics do not respect the fourfold logic of such names in the
manner required by Aristotle, Russell, Quine and ordinary language.

But there are even deeper, ontological consequences deriving from distinguishing logically proper names from ordinary proper names. For it becomes
evident that *individuals*, i.e. what the variables in such expressions as 'Kx'
vary over, must have eternal existence. So they must be separated from any
entities that merely have 'existence' in this world, or some other. What, indeed, in relation to *individuals*, has 'existence' just in this world, or just in
some other —making them 'physical objects', and 'fictions', respectively — are
identifying properties. To highlight the difference even more, we can say that
Aristotelian Realism holds for such physical objects/fictions, whereas *Platonic
Realism* holds for the associated individuals. It is more usual, perhaps, to think
of Platonic Realism and Aristotelian Realism as being rivals, in opposition to
one another, because they are seemingly differing accounts of the same thing.
But in the present connection we come to see that they are merely complementary, through being concerned with different things: identifying properties
that may or may not be instantiated, and the eternally existing individuals in
which those identifying properties (and other properties) may be instantiated.

The point is illustrated most clearly in the epsilon calculus theorem (which
will be proved in the next chapter, and see Slater 2009b, 417) showing that

$(\exists x)(Kx \ \& \ (y)(Ky \supset y=x) \ \& \ Bx)$

(i.e. 'A sole king of France exists, and is bald') is equivalent to

$(\exists x)(Kx \ \& \ (y)(Ky \supset y=x)) \ \& \ B\epsilon x(Kx \ \& \ (y)(Ky \supset y=x))$,

(i.e. 'A sole king of France exists. He is bald'). For the first conjunct in the
second expression is about certain identifying properties being instantiated.
That is what must hold for a sole king of France to exist (contingently). But
the second conjunct there is about a certain eternally existing individual —
one that is a sole king of France if there is such a thing, but which still exists
even if there is no such thing.

But not only through presenting Russell's formula as a conjunction do we enable a separation to be made between a true or false assertion about this world,
namely the first conjunct delimiting existence and uniqueness conditions, and a
further assertion, in the second conjunct, which is made about its subject independently of whether the first conjunct is true or false, and so about something
that exists eternally. For the same point is central to understanding how such
eternally real objects are accessed, which is a seemingly perennial difficulty
with Platonic entities. Paradigmatically the situation is represented again in
the epsilon variant to Russell's analysis of 'The king of France is bald'. Thus
while the above epsilon variant to Russell's

$(\exists x)(Kx \ \& \ (y)(Ky \supset y=x) \ \& \ Bx)$,

is

$(\exists x)(Kx \& (y)(Ky \supset y=x)) \& B\epsilon x(Kx \& (y)(Ky \supset y=x))$,

the first conjunct is itself equivalent to

$K\epsilon x(Kx \& (y)(Ky \supset y=x)) \& (y)(Ky \supset y=\epsilon x(Kx \& (z)(Kz \supset z=x)))$.

So access to the individual $\epsilon x(Kx \& (z)(Kz \supset z=x))$ is provided entirely by means of the linguistic act of supposing there is a sole king of France, and through its then being invariably possible to cross-refer to the same individual from within further assertions. Eternal objects, in this way, are simply subjects of discourse.

CHAPTER 3

COMPLETING RUSSELL'S LOGIC

1

The epsilon calculus improves upon the predicate calculus by systematically providing complete individual terms. Recent research has shown that epsilon terms are therefore the 'logically proper names' Russell was not able to formalise, but their use improves upon Russell's Theory of Descriptions not just in that way. This chapter details relevant formal aspects of the epsilon calculus before tracing its extensive application not just to the theory of descriptions, but also to more general problems with anaphoric reference.

2

In Russell's theory of definite descriptions there are, it will be remembered, three clauses: with 'The King of France is bald' these are 'there is a king of France', 'there is only one king of France' and 'he is bald'. Russell used an iota term to symbolise the definite description, but it is not an individual symbol: it is an 'incomplete' term, as he explained it, since 'The King of France is bald' is taken to have the complex analysis,

$(\exists x)(Kx \,\&\, (y)(Ky \supset y=x) \,\&\, Bx)$,

and so it does not have the elementary form 'Bx'. Russell hypothesised that, in addition to the linguistic expressions gaining formalisations by means of his iota terms, there was another, quite distinct class of expressions, which he called 'logically proper names'. Logically proper names would, amongst other things, take the place of the variable in such forms as 'Bx'. Russell suggested that demonstratives might be in this class, but he could give no further formal expression to them. Hilbert and Bernays, in their *Grundlagen der Mathematik*, introduce a kind of complete symbol, by contrast with Russell, defending what would later be called a 'presuppositional theory' of definite descriptions. The first two clauses of Russell's definition,

$(\exists x)(Kx \,\&\, (y)(Ky \supset y=x))$,

are not taken, by pre-suppositionalists, to be part of what is asserted by 'The King of France is bald'; they are, instead, the conditions under which one is allowed to introduce into the language an individual term for 'the King of France', which then satisfies the matrix of the quantificational expression above, and becomes a proper symbol to replace the variable in such expressions as 'Bx'. Hilbert and Bernays still used an iota term for this purpose, although it is quite different from Russell's iota term, since, when it is part of the language, it is equivalent to the related epsilon term. It has been realized, more recently, that epsilon terms, being complete symbols, are the 'logically proper names'

Russell was looking for, and that their natural reading is indeed as forms of demonstratives.

It is at the start of book 2 of the *Grundlagen* that Hilbert and Bernays introduce epsilon terms. They first go on to produce a theory of non-definite descriptions of the same pre-suppositional sort to their theory of definite descriptions. Thus they permit an eta term to be introduced into the language if the first of Russell's conditions is met, '$(\exists x)Kx$', this term then satisfies the associated matrix, but it is, in general, an individual, pre-suppositional term of the same kind as their iota one. There is a singular difference in certain cases, however, since the pre-supposition of the eta term can be proved conclusively, for certain matrices. Thus we know, for any predicate 'F', that

$$(\exists x)((\exists y)Fy \supset Fx),$$

since this is a theorem of the predicate calculus. The eta term this theorem permits us to introduce is what Hilbert and Bernays call an epsilon term. Thus we get the epsilon axiom

$$(\exists y)Fy \supset F\epsilon xFx,$$

which therefore implies

$$(\exists y)Fy \equiv F\epsilon xFx.$$

So an epsilon term is very unlike the generality of eta terms, since its introduction is clearly not dependent on any contingent facts about F. It is this that permits completely formal theories using epsilon terms to be developed, because such epsilon terms, unlike Hilbert and Bernays' iota terms, are always defined, and, as the equivalence indicates, they refer to exemplars of the property in question. The above predicate calculus theorem, in other words, provides the existence condition for certain objects, which the various epsilon calculi then go on to symbolise reference to, using epsilon terms. Copi has explained the theorem's relation with exemplars very fully (Copi 1973, 110).

Kneebone read epsilon terms as formalising indefinite descriptions (Kneebone 1963, 101), and this idea is commonly also found in the work of his pupil, Priest, although strangely Priest himself has pointed out that reading '$(\exists x)(Gx \ \& \ Fx)$' as '$G\epsilon xFx$' will not do (Priest 1979, 6), see also (Slater 1988, 285). Hilbert read the epsilon term in the above case 'the first F', which indicates its place in some, otherwise unspecified well-ordering of the F's - for instance, in connection with arithmetical predicates, that generated by the least number operator. So 'ϵxFx' is not 'an F'. Moreover, as Copi's discussion makes very clear, it is possible that an epsilon term refers to something which is in fact not F — it does this, of course, if there are no F's at all — and that will lead us to theories of reference which materialised only in the 1960s and later, when reference came to be properly distinguished from attribution. If there are F's then the first F is a chosen one of them; but if there are no F's then 'the first F' must be non-attributive, and so denotes something it cannot connote. It functions like a Millian name, in other words, with no applicable sense. With denotation in this way clearly distinguished from description we can then start to formalise the cross-reference that even Russell needed to link his first two conditions 'There is one and only one king of France' with his further condition 'He is bald'. For, by an extension of the epsilon equivalent of the existential condition, the 'he' in the latter comes to be a pronoun for the same epsilon term as arises in the former — whether or not the former is true. And such anaphoric cross-reference in fact may stretch into and across intensional contexts of the kind Russell was also concerned with, such as 'George IV wondered whether the author of *Waverley* was Scott'. For, of course, he was indeed Scott, and we may all now know very well that he was Scott. So we obtain a formalisation for transparency in such locutions.

That puts developed epsilon calculi at variance with Fregean views of intensional contexts — and also the Kripkean semantics which has continued to support Frege in this area. But Fregean intensional logic did not incorporate Millian symbols for individuals, and in particular, as we shall see in detail later, that meant it could not clearly distinguish individuals from their identifying properties. The addition of epsilon terms provides the facility for separating, for instance, 'Scott is the author of *Waverley*'

$$s = \epsilon x(y)(Ay \equiv y=x),$$

from 'Scott authored *Waverley*'

$$(y)(Ay \equiv y=s),$$

and so for isolating the proper object of George IV's thought.

3

When one begins to investigate the natural language meaning of epsilon terms, it is significant that Leisenring, writing in 1969, merely notes the 'formal superiority' of the epsilon calculus, comparing some of its pedagogic features with the comparable ones in the predicate calculus (Leisenring 1969, 63). Apparently its main value, in Leisenring's day, was that it could prove all that was provable in the predicate calculus but in a smarter, and less tedious way. Epsilon terms, for Leisenring, were just clever calculating instruments. Evidently there is more to the epsilon calculus than this, but until more recent times only the natural language meaning of the above epsilon axiom has been dwelt upon. There are a couple of further theorems within the epsilon calculus, however, which will show its extended range of application: they are about the nature and identity of individuals, as befits a calculus which systematically provides a means of reference to them.

The need to provide logically proper names for individuals only became generally evident some while after Russell's work on the theory of descriptions. The major difficulty with providing properly referential terms for individuals, in classical predicate logic, is what to do with 'non-denoting' terms, and Quine, following Frege, simply gave them an arbitrary, though specific referent. The approach was formalized perhaps most fully by Kalish and Montague, who gave the two rules (Kalish and Montague 1964, 242-3):

$$(\exists x)(y)(Fy \equiv y=x) \ / \ F\iota xFx,$$

$$\neg(\exists x)(y)(Fy \equiv y=x) \ / \ \iota xFx = \iota x(x{\neq}x),$$

where, in explicitly epsilon terms, we would have

$$\iota xFx = \epsilon x(y)(Fy \equiv y=x).$$

Kalish and Montague were of the opinion, however, that their second rule 'has no intuitive counterpart, simply because ordinary language shuns improper definite descriptions' (Kalish and Montague 1964, 244). And certainly, in that period, the revelations that Donnellan was to publish about non-attributive definite descriptions (Donnellan 1966) were not well known. But ordinary language does not, we now know, avoid non-attributive definite descriptions, although their referents are not as constant as Kalish and Montague's second rule requires. In fact, by being improper their referents are not fixed by semantics at all: like demonstratives the referents of logically proper names are found only in their pragmatic use. Stalnaker and Thomason were more appropriately liberal with their complete individual terms. And these referential terms also had to apply, they knew, in every possible world (Thomason and Stalnaker 1968, 363). But a fuller coverage of identity and descriptions, in modal and general intensional contexts, is to be found in (Routley, Meyer and Goddard 1974), and also (Hughes and Cresswell 1968). With these Australasian thinkers we find the explicit identification of definite descriptions with epsilon terms (e.g. Hughes and Cresswell 1968, 203).

Which further theorems in the epsilon calculus are behind these kinds of identification? There is one theorem in particular which demonstrates strikingly the relation between Russell's attributive, and some of Donnellan's non-attributive ideas (see Slater 1988). For

$(\exists x)(Fx \,\&\, (y)(Fy \supset y=x) \,\&\, Gx)$

is logically equivalent to

$(\exists x)(Fx \,\&\, (y)(Fy \supset y=x)) \,\&\, Ga,$

where $a = \epsilon x(Fx \,\&\, (y)(Fy \supset y=x))$. For the latter is equivalent to

Fa & (y)(Fy ⊃ y=a) & Ga,

which entails the former. But the former is

Fb & (y)(Fy ⊃ y=b) & Gb,

with $b = \epsilon x(Fx \,\&\, (y)(Fy \supset y=x) \,\&\, Gx)$, and so entails

$(\exists x)(Fx \,\&\, (y)(Fy \supset y=x)),$

and

Fa & (y)(Fy ⊃ y=a).

But then, from the uniqueness clause,

a = b,

and so

Ga,

making the former entail the latter.

The former expression, as we have seen, encapsulates Russell's Theory of Descriptions, in connection with 'The F is G'; it involves the explicit assertion of the first two clauses, to do with the existence and uniqueness of an F. A pre-suppositional account like that in Hilbert and Bernays, which was later popularised by Strawson, would not involve the direct assertion of these two clauses: on a pre-suppositional account they form the precondition without which 'the F' cannot be introduced into the language. But both of these accounts forget the use we have for non-attributive definite descriptions. Since Donnellan (and see Slater 1963), we now know that there are no preconditions on the introduction of 'the F'; and 'The F is G', as a result, may always be given a truth value. Hence 'Ga' properly formalises it. If the description is non-attributive, i.e. if the first two clauses of Russell's account are not both true, then the referent of 'The F' is simply up to the speaker to nominate.

But one detail about Donnellan's actual account must be noted at this point. He was originally concerned with definite descriptions which were improper in the sense that they did not uniquely describe what the speaker took to be their referent. And on that understanding the description might still be 'proper' in the above sense — if there still was something to which it uniquely applied. Specifically, Donnellan would originally allow 'the man with martini in his glass' to refer to someone without martini in his glass whether or not there was some unique man with martini in his glass. But someone talking about 'the man with martini in his glass' can be rightly taken to be talking about who this phrase describes, if it does in fact describe someone — Devitt and Bertolet pointed this out in criticism of Donnellan (Devitt 1974; Bertolet 1980). It is this latter part of our linguistic behaviour which the epsilon account of definite descriptions respects, for it permits definite descriptions to be referring terms without being attributive, but only so long as nothing has the description in question. Hence it is not the first quantified statement above, but only, so to speak, the third part of it extracted which makes the remark 'The F is G'.

This becomes plain when we translate the two statements using relative and personal pronouns:

There is one and only one F, which is G,

There is one and only one F; it is G.

For 'it' here is an anaphoric pronoun for 'the (one and only) F', and it still has this reference even if there is no such thing, because that is just a matter of the grammar of the language. Now the uniqueness clause is required for two such statements to be equivalent — without it there would be no equivalence, as we shall see — and that means that the relative pronoun 'which' is not itself equivalent to the personal pronoun 'it'. So it was because Russell's logic could not separate the (bound) relative pronoun from the (unbound) personal pronoun that it could not formulate the logically proper name for 'it', and instead had to take the whole of the first expression as the meaning of 'The F is G'. Using just the logic derived from Frege, it could not separate out the cross-referential last clause.

But how can something be the one and only F 'if there is no such thing'? This is where a second theorem in the epsilon calculus is relevant:

$$[Fc \mathbin{\&} (y)(Fy \supset y=c)] \supset [c = \epsilon x(Fx \mathbin{\&} (y)(Fy \supset y=x))].$$

For the singular thing is that this entailment cannot be reversed, so there is a difference between the left hand side and the right hand side, i.e. between something being alone F, and that thing being the one and only F. We get from the left hand side to the right hand side once we see the left hand side entails

$$(\exists x)(Fx \mathbin{\&} (y)(Fy \supset y=x)),$$

and so

$$F\epsilon x(Fx \mathbin{\&} (y)(Fy \supset y=x)) \mathbin{\&} (z)(Fz \supset z=\epsilon x(Fx \mathbin{\&} (y)(Fy \supset y=x))).$$

By the uniqueness clause, and Fc, we get the right hand side. But if we substitute '$\epsilon x(Fx \mathbin{\&} (y)(Fy \supset y=x))$' for 'c' in the whole implication then the right hand side is necessarily true. But the left hand side is then equivalent to

$$(\exists x)(Fx \mathbin{\&} (y)(Fy \supset y=x)),$$

which is, in general, contingent; hence the implication cannot be logically reversed.

The difference is not available in Russell's logic. In fact Russell confused the two forms, since possession of an identifying property he formalised using the identity sign

c = ιxFx,

making it appear that some, maybe even all identities are contingent. But all proper identities are necessary, and it is merely associated identifying properties that are contingent. Ironically, Frege used a complete term for definite descriptions in his extensional logic, as was mentioned before. But Russell explicitly argued against the arbitrariness of Frege's definition, in the case where there isn't just one F, when setting up his alternative, attributive theory of descriptions, in 'On Denoting'. Had Frege's complete term been more widely used, and, for a start, been used in his intensional logic, results like those above might have been better known earlier.

Hughes and Cresswell, at least, appreciated that in addition to 'contingent identities' there were also necessary identities, and differentiated between them as follows (Hughes and Cresswell 1968, 191):

> Now it is contingent that the man who is in fact the man who lives next door is the man who lives next door, for he might have lived somewhere else; that is *living next door* is a property which belongs contingently, not necessarily, to the man to whom it does belong. And similarly, it is contingent that the man who is in fact the mayor is the mayor; for someone else might have been elected instead. But if we understand ['The man who lives next door is the mayor'] to mean that the object which (as a matter of contingent fact) possesses the property of being the man who lives next door is identical with the object which (as a matter of contingent fact) possesses the property of being the mayor, then we are understanding it to assert that a certain object (variously described) is identical with itself, and this we need have no qualms about regarding as a necessary truth. This would give us a way of construing identity statements which makes [(x=y) ⊃ L(x=y)] perfectly acceptable: for whenever x=y is true we can take it as expressing the necessary truth that a certain object is identical with itself.

There is more hangs on this matter, however, than Hughes and Cresswell appreciated. For now that we have the logically proper names, i.e. complete symbols to take the place of the variables in such expressions as 'x=y', not only do we see better where the contingency of the properties of such individuals comes from - just the linguistic possibility of improper definite descriptions - we also see, contrariwise, why constant epsilon terms must be rigid — because identities involving such terms are necessary.

Frege, for instance, thought that we could not derive 'a believes the Morning Star is illuminated by the sun' from 'a believes the Evening Star is illuminated by the sun', even though the Morning Star is the Evening Star. But (see, for instance, Slater 1992a), from

Ba$\iota\epsilon$xEx

we can derive

BaIϵxMx,

if ϵxEx=ϵxMx; what we cannot derive is

Ba(∃x)[(y)(My ≡ y=x) & Ix]

from

Ba(∃x)[(y)(Ey ≡ y=x) & Ix],

even if

(∃x)(∃y)(Mx & Ey) & (x)(y)[(Mx & Ey) ⊃ x=y].

Russell improved matters somewhat, by distinguishing a primary, transparent sense

(∃x)[(y)(Ey ≡ y=x) & BaIx]

from the secondary, opaque sense

Ba(∃x)[(y)(Ey ≡ y=x) & Ix],

since the former, with

(∃x)(∃y)(Mx & Ey) & (x)(y)[(Mx & Ey) ⊃ x=y],

does entail

(∃x)[(y)(My ≡ y=x) & BaIx].

But without epsilon terms to provide explicit instantiations of the primary-sense forms, Russell was in no position to detach their second conjuncts. With that facility we can write, for instance, the last expression as a conjunction

(y)(My ≡ y=b) & BaIb,

with b = ϵx[(y)(My ≡ y=x) & BaIx], and so read it as

b is alone a morning star, and a believes it is illuminated by the sun.

We thereby provide transparent access to the individual in the intensional construction, since the 'it' cross-refers to the (extensional) object in the antecedent clause before.

Some grammarians have tried to handle this sort of issue in intensional contexts by returning to Meinongian 'intensional objects', or the 'counterparts' of actual individuals in alternative worlds. For example, Saarinen considers the following case (Saarinen 1978, 277):

> Bill believes that the lady on the stairs [is acquainted with] him,
> but John knows she is only a wax figure.

About this Saarinen says 'Both of the attitudes are of the wax lady, and yet all the relevant individuals in the doxastic worlds are not wax ladies but human beings' (Saarinen 1978, 282). However, in addition to human beings in this world it is not that there are also human beings in people's minds, merely, also, that there are things in this world which are taken to be human beings, by people. Saarinen supports his judgement with a Russellian, i.e. attributive reading of 'the lady on the stairs', but more important is his retention of the Meinongian idea that such an intensional object as the gold mountain has to be made of gold, and even that an impossible intensional object like the round square has to still be both round and square. Thereby, of course, Saarinen misses the possibility that what Bill believes is the one and only lady on the stairs is not really a lady on the stairs. And it is this possibility, exactly, which allows it to be just the plain, and everyday, physical wax figure which is the object of both Bill's and John's attitudes. The form of Saarinen's case, if 'the lady on the stairs' is attributive, is

 Bb($\exists$x)((y)(Ly $\equiv$ y=x) & Abx) & KjWϵxBb((y)(Ly $\equiv$ y=x) & Abx),

but the second conjunct of this entails

 $\neg$LϵxBb((y)(Ly $\equiv$ y=x) & Abx),

since knowledge entails truth, and being a wax figure entails not being a lady.

 As was mentioned before, it is even contingent that *the lady on the stairs* is a lady on the stairs, but the source of this non-doxastic, and simply modal contingency, which allows the same object to appear in other possible worlds, cannot be properly seen until we link it with the linguistic possibility of improper, i.e. non-attributive definite descriptions. Seeing the source of this even more radical contingency is thus essentially linked to seeing how there can be *de re* attitudes. But it is also directly linked with the much more substantial programme of replacing such a metaphysical view as Meinong's simply with accurate linguistic analysis: 'philosophical problems arise through misconceptions of grammar' said Wittgenstein.

4

It follows that there is no essential grammatical difference between such an intensional anaphoric remark about someone's mind, as

> The ancients believed there was a star in the morning that was illuminated by the sun. But it was a planet.

i.e.

 Ba($\exists$x)(Mx & Ix) & PϵxBa(Mx & Ix),

and the extensional cross reference, for instance, in

 There was a man in the room. He was hungry.

i.e.

 ($\exists$x)Mx & HϵxMx.

What has been the problem, fundamentally, has been getting the cross-reference formalised first of all in the purely extensional kind of case. Yet this just requires extending the epsilon replacement for an existential statement, by means of a repetition of the associated epsilon term, as was mentioned with respect to 'he' in Russell's case, before. The only difference in the intensional case is that, to obtain the required cross-referencing one must move from

 Ba($\exists$x)(Mx & Ix)

to

 ($\exists$x)Ba(Mx & Ix)

via

 Ba(Mb & Ib)

with b=ϵx(Mx & Ix) to get a public referential phrase for the object. And note that, while the required epsilon term 'ϵxBa(Mx & Ix)' is then defined intensionally, it still refers to a straightforward extensional object — the planet Venus, of course.

 It is now better understood how the epsilon calculus allows us to do this (see Slater 1986, Purdy 1994, Egli and von Heusinger 1995, Meyer Viol 1995, Ch. 6, for instance). The starting point is the possibility illustrated in the theorem about Russellian definite descriptions before, of separating out what otherwise, in the predicate calculus, would be a single sentence into a two-sentence piece of discourse, leaving the existence and uniqueness clauses in one place, and putting the characterising remark in another. The point really starts to matter when there is no way to symbolise in the predicate calculus some anaphorically linked remarks where there is no uniqueness clause, as in the above extensional case. This is what became a problem for the Fregean and Russellian logicians who woke up to the need to formalize anaphoric reference in the 1960s.

It can be seen, in retrospect, how it was lack of the epsilon calculus that was the major cause of the difficulty. Thus Geach, in an early discussion of the issue, went to the extremity of insisting that there could be no syllogism of the following form (Geach 1962, 126):

> A man has just drunk a pint of sulphuric acid.
>
> Nobody who drinks a pint of sulphuric acid lives through the day.
>
> So, he won't live through the day.

Instead, Geach said, there was only the existential conclusion:

> Some man who has just drunk a pint of sulphuric acid won't live through the day.

Certainly one can only conclude
$$(\exists x)(Mx \ \& \ Dx \ \& \ \neg Lx)$$
from
$$(\exists x)(Mx \ \& \ Dx),$$
and
$$(x)(Dx \supset \neg Lx),$$
within Fregean predicate logic. But one can still conclude
$$\neg L\epsilon x(Mx \ \& \ Dx),$$
within its conservative extension: Hilbert's epsilon calculus.

And through inattention to that extension, Geach was entirely stumped later, in (Geach 1967), when he discussed his famous intensional example (3)

> Hob thinks a witch has blighted Bob's mare, and Nob wonders whether she (the same witch) killed Cob's sow,

i.e.
$$Th(\exists x)(Wx \ \& \ Bxb) \ \& \ OnK\epsilon xTh(Wx \ \& \ Bxb)c.$$
For he saw this could not be his (4)
$$(\exists x)(Wx \ \& \ ThBxb \ \& \ OnKxc),$$
or his (5)
$$(\exists x)(Th(Wx \ \& \ Bxb) \ \& \ OnKxc).$$
But a reading of the second clause as
   Nob wonders whether the witch who blighted Bob's mare killed Cob's sow (c.f. Geach's 18) in which 'the witch who blighted Bob's mare killed Cob's sow' is analysed in the Russellian manner, as Geach's (20):

> just one witch blighted Bob's mare and she killed Cob's sow,

does not provide the required cross-reference — for one thing because of the uniqueness clause then involved. Of course the descriptive replacement for the personal pronoun 'she' in the Hilbertian expression, namely 'what Hob thinks is a witch that blighted Bob's mare', does not have any implication of uniqueness.

The inappropriateness of the uniqueness clause in Russellian analyses has been widely discussed. However it did not deter Neale who much later wrote a whole book defending a largely Russellian account of definite descriptions, and cross-sentential anaphora. But he got no further with Geach's case above than proposing that 'she' might be localised to 'the witch we have been hearing about' (Neale 1990, 221), thinking in general that definite descriptions merely should be relativised to the context. But a greater change is needed than that. For, as before, it is not that, in addition to witches in the actual world, there are also witches in people's minds, but merely that, in addition to witches in the actual world there are things in the actual world which are thought, or believed to be witches. Geach's 'the same witch' is also inappropriate, on the same grounds.

A large amount of the important, initial work in this area was done by another person very influenced by the Russellian tradition: Evans. But Evans also explicitly separated from Russell over the matter of uniqueness, for instance in connection with back-reference to a story about a man and a boy walking along a road one day (Evans 1977, 516-7):

> One does not want to be committed, by this way of telling the story, to the existence of a day on which just one man and boy walked along a road. It was with this possibility in mind that I stated the requirement for the appropriate use of an E-type pronoun in terms of having answered, or *being prepared to answer upon demand*, the question 'He? Who?' or 'It? Which?' In order to effect this liberalisation we should allow the reference of the E-type pronoun to be fixed not only by predicative material explicitly in the antecedent clause, but also by material which the speaker supplies upon demand. This ruling has the effect of making the truth conditions of such remarks somewhat indeterminate; a determinate proposition will have been put forward only when the demand has been made and the material supplied.

It was Evans who popularised the name 'E-type pronoun' for the pronoun in such cases as

A Cambridge philosopher smoked a pipe, and he drank a lot of whisky,

i.e.

$(\exists x)(Cx \ \& \ Px) \ \& \ D\epsilon x(Cx \ \& \ Px).$

He also argued at length, in line with the above (Evans 1977, 516), that what was distinctive about E-type pronouns was that such a conjunction of statements as this was not equivalent to

A Cambridge philosopher, who smoked a pipe, drank a lot of whisky,

i.e.

$(\exists x)(Cx \ \& \ Px \ \& \ Dx).$

Obviously the epsilon account supports this, since the contrast illustrates the point remarked before: only the expression which contains the relative pronoun can be symbolised in the predicate calculus, since to symbolise the personal pronoun its epsilon extension is needed.

The extreme subtlety of expression that is then available can be illustrated in what have sometimes been called 'donkey anaphora'. For instance, how is one to analyse the conditional 'If a man gets a new car, he washes it'? The epsilon analysis is:

$(\exists x)(\exists y)(Mx \& Cy \& Gxy) \supset Wab$,

where $a = \epsilon x(\exists y)(Mx \& Cy \& Gxy)$, and $b = \epsilon y(Ma \& Cy \& Gay)$. But the inclination, from a Fregean perspective, is to read the conditional as a universal expression, 'Every man who gets a new car washes it', i.e.

$(x)(y)[(Mx \& Cy \& Gxy) \supset Wxy]$.

The problem with this analysis, however, is that it is an analysis of an allied but quite distinct form, namely

Invariably if a man gets a new car he washes it.

For we can also have, for instance,

Frequently if a man gets a new car he washes it,

and

Rarely if a man gets a new car he washes it,

and even,

Never if a man gets a new car he washes it.

These further examples show that the original conditional is one of a kind that was never considered in the times when Fregean logic was dominant. In fact it is a conditional whose existence has been denied: it is a conditional whose probability is a conditional probability, making it satisfy Adams' and Stalnaker's Hypothesis. For if there is a man who gets a new car then the probability of the original conditional is the chance that the 'he' and the 'it' picked out are amongst those pairs such that the one washes the other. So it is simply the proportion of men who get cars who wash them, i.e. the conditional probability in question. If there is no man who gets a new car then the probability of the antecedent of the conditional is zero, making the probability of the conditional 1, which we can arbitrarily stipulate to be the value of the conditional probability in such a case. But if there is a man who gets a new car then the two epsilon terms '$\epsilon x(\exists y)(Mx \& Cy \& Gxy)$', and '$\epsilon y(Ma \& Cy \& Gay)$' pick appropriate men and cars, the first a man who gets a car, the second the car that he gets.

5

I have presented many other examples of the subtlety of the Epsilon Calculus in other works. Thus another feature of such conditionals as those immediately above (c.f. Slater 2004c) is that there is no difficulty with using the material conditional for 'if', even when the case is subjunctive, thus preserving the natural grammar of, for instance, 'If there were chickens there would be eggs'. In this case the modification needed to express the conditional lies simply in the formalisation of its antecedent and consequent in terms of the appropriate world. Writing '$V[p, i] = 1$' for 'it is true in world i that p', the conditional

If a world was to contain chickens, then it would contain eggs,

can be formalized:

$$(\exists i)(V[c, i] = 1) \supset V[e, \epsilon i(V[c, i] = 1)] = 1,$$

with 'c' being 'there are chickens' and 'e' being 'there are eggs'. Here the cross-sentential anaphor 'it' is symbolized by the epsilon term generated by the antecedent clause, and so refers to a chosen world in which there are chickens if there are any, and an arbitrary world, if not. Again the conditional can be qualified by modifiers estimating how likely the consequent is, on the given supposition, and so is a conditional whose probability is a conditional probability. Epsilon terms, in this place have some similarities with the selection functions in Stalnaker's theory of conditionals, but Stalnaker, amongst other things, assumes an 'absurd world' in which everything is true to handle the case where the antecedent is false, and that is avoided here through the stipulation that the conditional probability is 1. The above paper also compares the epsilon account of such conditionals to a related probabilistic theory of causation by Judea Pearl.

Again (c.f. Slater 2005b), there are referential paradoxes which are immediately resolved once one recognises the viability of epsilon terms. One immediate consequence concerns interpretations of natural language generating Berry's Paradox, and the like. Is there a least number not referable to in less than 19 syllables? The phrase 'the least number not referable to in less than 19 syllables' contains less than 19 syllables, and so must be non-attributive in its application, which means it is arbitrary in its reference. But that means that, only functionally dependent on the consequent arbitrary choice can one determine a set of numbers referable to in less than 19 syllables, and so a least number in the then complementary set. There is no constant set, or least number not in it available, which is why calling any number 'the least number not in the set of numbers referable to in less than 19 syllables' must be a context-sensitive, pragmatic matter.

But also other paradoxes have been invented which are resolved on the same principles. Suppose, for instance, one puts three referential expressions on the board 'six', 'π', and 'the sum of the numbers denoted by expressions on this board'. One might be tempted to conclude, first, that, since $\neg(\exists n)(n = 6 + \pi + n)$ the third phrase did not denote anything. But then, secondly, that if it does not denote anything, then the sum of the numbers denoted by expressions on the board is $6 + \pi$, which means that the third phrase does denote something, after all. The case is a particular problem for pre-suppositional theories of definite descriptions. For adherents of such theories would immediately jump to the first conclusion, saying that, in the absence of the required existence, the third phrase is fictional, and does not denote. Yet the remaining expressions clearly do denote, and denote numbers moreover, so there is a sum of the numbers denoted, making the phrase 'the sum of the numbers denoted by expressions on this board', contradictorily, non-fictional. If one is tempted to argue in this kind of way, however, it is merely because one has been schooled to choose interpretations of such cases that lead to paradox, when there are other interpretations which generate no trouble. For even if $\neg(\exists n)(n = 6 + \pi + n)$, still the third phrase may have a reference — simply a non-attributive one.

If a = ϵn(n = 6 + π + n), then, because ($\exists$x)Fx $\equiv$ FϵxFx, by the definition of epsilon terms, all that follows is that $\neg$(a = 6 + π + a), and so 'a' can be chosen at will.

The second part of the above case also has an alternative, counterfactual reading, since it involves the anaphoric expression 'then'. If the third phrase on the board indeed *had not* referred to anything, the sum of the numbers *then* denoted by expressions on the board would be 6 + π. But the referring phrase just used is different from the third phrase on the board, because of the presence of the 'then'. So it is not that there is one expression that does and doesn't denote something, there are, on this reading, simply two different expressions, one referring to what is on the board in the actual world, and the other referring to what would be on the board in the further, imagined circumstances. In the actual case, of course, there is simply a choice about what the original phrase refers to, as with the phrase in Berry's Paradox. So, if the original phrase is chosen to refer to n, the sum emphthen is simply 6 + π + n, which means there is a sum of the numbers but it is non-constant, since it is functional on another variable: (n)($\exists$m)(m = 6 + π + n). The variability of the sum explains, in a different way, why the third expression on the board, in actual fact, must be non-attributive.

CHAPTER 4

ONTOLOGICAL DISCRIMINATIONS

1

Russell held that 'a exists', where 'a' is a logically proper name, was necessarily true. By contrast his account of 'The K exists' allowed this to be contingent, since, on his Theory of Descriptions, it did not assert the existence of an individual, but merely the instantiation of some uniquely identifying properties. The present chapter refines Russell's distinction in several ways, first by providing what Russell merely gestured at, namely explicit, formally defined logically proper names. But following from this it is seen that Russell's intention with regard to 'The K exists' is better expressed 'A unique K exists', leaving the former to be assimilated into the non-contingent category, through interpreting its subject phrase 'The K' non-attributively. The chapter closes with an exhibition of similar discriminations that are available with higher-order subjects, such as properties, numbers, and facts; and most particularly with propositions.

2

We must first have clear what referential terms are. This was obscured in Russell's Theory of Descriptions, as we have seen, and remains unclarified to this day because of the popularity of that theory, and the associated neglect of Hilbert's Epsilon Calculus. In Russell's Theory there are, it will be remembered, three clauses with 'The King of France is bald'. These are 'there is a king of France', 'there is only one king of France' and 'he is bald'. Russell used an iota term to symbolise the definite description, but it is not an individual symbol: it is an 'incomplete' term, as he explained it, since 'The King of France is bald' is taken to have the complex analysis,

$(\exists x)(Kx \ \& \ (y)(Ky \supset y = x) \ \& \ Bx)$,

and so it does not have the elementary form 'Bx'. Russell hypothesised that, in addition to the linguistic expressions gaining formalisations by means of his iota terms, there was another, quite distinct class of expressions, which he called 'logically proper names'. Logically proper names would, amongst other things, take the place of the variable in such forms as 'Bx'. Russell suggested that demonstratives might be in this class, but he could give no further formal expression to them. The epsilon theory of descriptions that settles the question was discussed in the first edition of Hughes and Cresswell's classic introductory text on Modal Logic, and originated with Routley, Meyer and Goddard, who, in their work on intensional contexts, made an explicit identification of definite descriptions with epsilon terms: The King of France = $\epsilon x(Kx \ \& \ (y)(Ky \supset y = x))$, (Goddard and Routley 1973, 558; Hughes and Cresswell 1968, 203).

We saw before, in chapters 2 and 3, which theorems in the epsilon calculus

are behind this kind of identification. The standard epsilon calculus contains
the axiom '$(\exists x)Fx \supset F\epsilon xFx$' (Leisenring 1969, Meyer Viol 1995), from which
one can naturally obtain the equivalence between the two sides. There is then
one theorem in particular which demonstrates strikingly the relation between
Russell's attributive, and Donnellan's 'purely referential' understanding of referential terms. For

$(\exists x)(Kx \ \& \ (y)(Ky \supset y = x) \ \& \ Bx)$,

is logically equivalent to

$(\exists x)(Kx \ \& \ (y)(Ky \supset y = x)) \ \& \ Ba$,

where $a = \epsilon x(Kx \ \& \ (y)(Ky \supset y = x))$, this being the epsilon term arising from
the first conjunct in the second expression. The first expression, as we have
seen, encapsulates Russell's Theory of Descriptions, in connection with 'The K
is B'; it involves the explicit assertion of the first two clauses, to do with the
existence and uniqueness of an K. Since Donnellan, however (Donnellan 1966),
we have realized that there are no preconditions on the introduction of 'the K'
as an individual term. So 'The K is B', with 'The K' an individual term, may
always be given a truth value, even if, sometimes, that truth value is merely an
arbitrarily chosen one. 'Ba' properly formalises 'The K is B', since the cross-
reference in the second expression means that it reads 'There is a single K. It is
B', and the descriptive replacement for the E-type pronoun 'it', there, is 'The
K'. If the description in 'a' is non-attributive, i.e. if the first two clauses of
Russell's account are not both true, and there is no such thing as the King of
France, then the referent of 'the K' is simply up to the speaker to nominate -
perhaps as a private, or secret name for something.

How can something be the one and only K 'if there is no such thing'? That
is where the second theorem proved before is even more important:

$(Ka \ \& \ (y)(Ky \supset y = a)) \supset [a = \epsilon x(Kx \ \& \ (y)(Ky \supset y = x))]$.

For the singular thing is that this entailment cannot be reversed, so there is
a difference between the left hand side and the right hand side, i.e. between
something being alone king of France, and that thing being the one and only
king of France. The difference is not available in Russell's logic, as we saw since
only possession of the property can be formalised there. In fact Russell confused
the two forms, since possession of an identifying property he formalised using
the identity sign, viz '$a = \iota x Kx$', making it appear that some, maybe even
all identities are contingent. But all proper identities are necessary, and it
is merely associated identifying properties that are contingent. That means
that in all possible worlds there is the same domain of discourse, although
the individuals in that domain may change their properties, and even their
individuating properties, from one world to the next. One specific form of this
distinction is of particular moment in what follows: the distinction between
possession of the same name, and being the same individual. We shall find
that traditional modal semantics has confused the two, and that only their
separation can give a clearer picture of identity in modal contexts.

3

What are the consequences of the above for modal, and general intensional
logics? Clearly, if the same individual is to be involved, epsilon terms must

be rigid across all worlds. Richard Routley presented several rigid intensional semantics, some objectival and some substitutional (Routley 1977, 185186). One of these semantics, for instance, simply took the first epsilon axiom to hold in any interpretation, and made the value of any epsilon term itself. Using such a rigid semantics, Routley, Meyer and Goddard obtained what has been called 'Routley's Formula', i.e.

L(∃x)Fx ⊃ (∃x)LFx,

by means of the following transformations. Routley's Formula holds for any propositional operator and any predicate, but they illustrated it in the case of necessity, and with 'Fx' as 'x numbers the planets'. Then, with 'εxFx' as 'the number of the planets', they said that from 'L(∃x)Fx', we can get 'LFεxFx', by the epsilon definition of the existential quantifier, and so '(∃x)LFx', by existential generalisation over the rigid term (Routley, Meyer and Goddard 1974, 308; Hughes and Cresswell 1968, 197, 204). (For further discussion of Routley's Formula see, for instance, Slater 1992b).

We therefore see not only that L(∃x)(x=a), because it is provable that some-thing is a, but also that (∃x)L(x=a), i.e. that the same thing is a in all pos-sible worlds. In Kripkean semantics it seems we can discriminate between 'L(∃x)(x=a)' and '(∃x)L(x=a)' on the grounds that the former merely says that in every possible world there is something named 'a', while the latter says that the same thing is named 'a' in every possible world. But this way of look-ing at the formulae, we now see, involves a use-mention confusion. Certainly one can discriminate between 'in every possible world something is named 'a'', and 'the same thing is named 'a' in every possible world', but these are meta-linguistic, relational remarks, which ought to be symbolised '(y)(∃x)Nx'a'y', and '(∃x)(y)Nx'a'y', where 'Nx'a'y' says that x is named 'a' in world y.

In contrast to these facts about mentioned terms, we saw before that the main consequence of the rigidity of used terms is that individuals, properly so-called, have eternal existence, and so must be separated from any entities that merely have 'existence' in this world, or some other (see also Stalnaker 2003, 120-126). The point was also central to understanding how eternally real objects are accessed: access to the object εx(Kx & (z)(Kz ⊃ z=x)), for instance, is provided entirely by means of the linguistic act of supposing there is a sole king of France, and through its then being invariably possible to cross-refer to the same individual from within further assertions. In the final chapter in the body of the book we will see more explicitly how all of this supports very directly the thesis that logic is not mathematical.

But the above merely presents the required details in the case of first-order subjects, while, by analogy with it, parallel things hold with higher-order sub-jects. So we must now also show how the commonly supposed abstraction of higher-order subjects, like properties, facts, and numbers has to be re-thought in the light of the foregoing. For if someone is happy, then Happiness in a cer-tain sense exists in the physical world, just like any other physical thing — and likewise with the fact that someone is happy, and with the number of people that are happy — for instance, the 10 people that are happy. But if no-one is happy then Happiness can, in a clear sense, only be dreamt about, and both the fact that someone is happy and the ten people that are happy are fictions.

So whether properties, facts, and numbers of things are 'abstract' or 'real' objects might also be said to be contingent. Following Cocchiarella (Cocchiarella 1986) I shall formalise the property of being happy as $\lambda y Hy$, the property of being 10 in number as $\lambda P(10y)Py$, the proposition that someone is happy as $\lambda(\exists z)Hz$, and so the fact that someone is happy as $\epsilon y(Ty \ \& \ y=\lambda(\exists z)Hz)$.

Then it is contingent whether $(\exists x)Hx$, but not whether $(\exists x)(x=\lambda y Hy)$, and it is contingent whether $(10x)Hx$, but not contingent whether $(\exists x)(x=\lambda P(10y)Py)$. Also it is not contingent whether $(\exists x)(x=\epsilon y(Ty \ \& \ y=\lambda(\exists z)Hz))$, or contingent whether $(\exists x)(x=\lambda(\exists y)Hy)$, but it is contingent whether $(\exists y)(Ty \ \& \ y=\lambda(\exists x)Hx)$ $(\equiv T\lambda(\exists x)Hx, \equiv (\exists x)Hx)$. Of course, if there is a property, a natural number, or a fact, that does not arise in the actual world in these ways, then it would be counted as unreal in the Aristotelian sense, and so could be said to be 'abstract' or 'fictional'. But always, with respect to the right kind of object, associated with Platonic Realism clearly, there is no contingency. And with each such object access to it is merely through allusion to it, in the appropriate language. Thus the fact that someone is happy, i.e. $\epsilon y(Ty \ \& \ y=\lambda(\exists z)Hz)$ $(=h)$, is alluded to when it is claimed that it is true that someone is happy i.e. that $(\exists y)(Ty \ \& \ y=\lambda(\exists x)Hx)$ $(\equiv (Th \ \& \ h=\lambda(\exists x)Hx))$. But the fact exists to be talked about even if it is not true that someone is happy.

4

More dramatically, with propositional referential terms in hand we can now not only symbolise many forms of speech which have been overlooked since Quine and before, but also validate many forms of inference that have been hidden from view during the same period. (In fact we can also resolve all paradoxes in the Liar family, as is shown in chapters 4, 5 and 6).

For one quite new form of expression comes into logical theory as a result of having referring terms to propositions: propositional identities of the form 'x = λp'. There are many such with names in place of 'x' — for instance those explicatiing the substance of Pythagoras' Theorem, Goldbach's Conjecture, The Peter Principle, and Murphy's Law. Quite commonly, however, the present generation of logical theorists has tried to insert a variety of set-theoretic expressions in place of 'x'. But propositions are not sets of any kind, since propositions are the referents of nominalised sentences. A nominalising functor is needed, therefore, and the one involving the extension of the Lambda Calculus that Cocchiarella provided recognizes most clearly the true continuity with concepts developed in the past. Using this new language it is then easy to symbolise, and validate many intensional inferences in natural language that previously have been left unformalised. Thus from

My proposal is that we go to Benidorm for our holiday,
(i.e. 'm = λp') with

You accept my proposal,
(i.e. ' Aym'), we can get

You accept that we go to Benidorm for our holiday,
(i.e. 'Ayλp'). It is significant that this inference cannot be validated using an operator expression for 'you accept that', since then nothing of the form 'Aym' or 'Ayλp' would be available in which substitution of the original identity could

take place. An operator expression, such as 'Oyp', would not discriminate a place for a nominal expression referring to a proposition, like 'm' or 'λp', since it would fuse the associated nominaliser 'that' with the verb.

But there are also quantificational discriminations between operator forms and referential forms that are highly significant. For we can now symbolise, first of all, *entailments*: 'A $\to$ B' is not 'A entails B' (for any '$\to$'), since 'entails' is a verb, and so needs nominal expressions on either side of it; properly it is 'that A entails that B', i.e. '$\lambda AE\lambda B$' ($\equiv L(A \supset B)$). That is not just a minor grammatical matter, however, since without the nominalising facility provided by 'that'-clauses, and similar referential phrases, there are major difficulties when one turns to modal logic. Thus, in that area, many might follow Prior (c.f. Leitgeb 2009), who did not use such referential constructions, and say that it was provable that

$(\exists p)([L(p \supset B) \lor L(p \supset \neg B)] \,\&\, p)$.

For instances of its matrix (call it 'Cp') seemingly follow from B, and also from $\neg B$. But if there are propositional epsilon terms in this style, i.e. if $(\exists p)Cp \equiv C\epsilon pCp$, then we could prove that

$[L(\epsilon pCp \supset B) \lor L(\epsilon pCp \supset \neg B)] \,\&\, \epsilon pCp$,

and so, by Necessitation, that

$L\epsilon pCp$,

and by rule K, that

$(L\epsilon pCp \supset LB) \lor (L\epsilon pCp \supset L\neg B)$.

Hence we would get that $LB \lor L\neg B$ (for any 'B'), and so modal collapse. But if one uses referential propositional quantification, no such conclusion can be drawn. For then 'LB' can be represented 'NλB', with 'N' the predicate of 'that'-clauses and similar nominals, 'is necessary'. Likewise 'Tr' says that r is true, with $T\lambda A \equiv A$, and rule K becomes: if rEs then Nr *supset* Ns. Certainly then we can say that there is a true proposition which either entails that B or entails that $\neg B$, i.e.

$(\exists r)([rE\lambda B \lor rE\lambda \neg B] \,\&\, Tr)$.

But if the matrix there is 'Dr' then, in the referential epsilon equivalent, we only get that

$[\epsilon rDrE\lambda B \lor \epsilon rDrE\lambda \neg B] \,\&\, T\epsilon rDr$,

and so that

$LT\epsilon rDr$,

and that

$(N\epsilon rDr \supset N\lambda B) \lor (N\epsilon rDr \supset N\lambda \neg B)$.

So $N\lambda B \lor N\lambda neg B$ is not obtainable.

If ϵrDr was necessary ($N\epsilon rDr$) we could again draw the conclusion that there was modal collapse, but all that is necessary is that ϵrDr is true ($LT\epsilon rDr$, i.e. $N\lambda T\epsilon rDr$), which allows ϵrDr to be contingent. In understanding this difference it is important to remember that the referential epsilon term is not itself an expressed proposition, but merely a pro-form referring to the proposition alluded to in the existentially quantified form before.

CHAPTER 5

OUT OF THE LIAR TANGLE

1

There are some seemingly small points to be made, first of all, about use-mention confusions in Stephen Read's paper 'The Truth Schema and the Liar' (Read 2008a). But underlying them is a grammatical point that has much wider repercussions. For it generates, on its own, a more straightforward way of understanding what gets people into a tangle with Liar and Strengthened Liar sentences, and that leads to a much fuller, critical assessment of the line of approach to these matters that Read derives from Bradwardine. Difficulties with representing propositional referring phrases of the form 'that p' are shown to be what have made Liar and Strengthened Liar sentences seem paradoxical. Using " 'p' " as an alternative confuses syntactic expressions with their immediate readings, and leads to misunderstandings about the necessity of the T-schema. Frege's content stroke, i.e. the horizontal line that he used to indicate the thought expressed by a sentence, has not been incorporated into the generality of logic texts that have followed his formal work. I set out here how provision of an explicit content marker gets one out of Liar and Strengthened Liar 'paradoxes', and what similarities and differences there are with Read's account of truth, as a result.

2

When discussing Tarski's T-schema, at the beginning of his paper, Read also mentions a similar schema in Horwich's work. He even thinks of them as the same, saying at one point 'Tarski did not propose (T) as a definition of truth, though others, e.g. Horwich, have done so since'. But there is a crucial difference between Tarski's and Horwich's schemas. Horwich's is a propositional truth schema, viz 'the proposition that p is true if and only if p', whereas Tarski's is a sentential schema: x is true if and only if p, when what replaces 'x' is a name of a sentence whose translation into the metalanguage replaces 'p'. The difference is most pointed in the homophonic sentential case, which parallels very closely the propositional one. For what replaces 'x' then is a quotation-name of the sentence that replaces 'p', not that sentence itself. As we shall see, there is a formal difficulty with distinguishing the two schemas clearly, since there is no agreed, distinct symbolisation for predications of the form 'that p is true', and something more like the sentential predication 'p' is true is often used in place of it. Certainly there would be no need to make the distinction if all sentences were unambiguous and non-indexical, i.e. had just one interpretation, since then facts about propositions could be mapped 1-1 onto facts about sentences. But, as we shall see later, the central question is whether Liar sentences are indeed univocal in the required way.

It has to be said that the difference between 'that p is true', and " 'p' is true" may not be completely appreciated even in Horwich's informal work, since he thinks there are still paradoxical cases of his propositional schema. But 'it is true that' is the null, or vacuous modality in the modal system KT, i.e. an 'L' for which it is necessary that Lp ≡ p, and so one cannot have p ≡ ¬Lp, since KT is consistent. Going with that, there is no way to construct self-referential propositions parallel to sentential constructions that, for instance, produce identities like

   t = 't is not true'.

For 't' is there the name of the sentence, and not the sentence itself, and one cannot have anything of the form

   'p' = 'that p is not true',

since nothing can be a proper part of itself. This shift in grammar already is revealing about Read's core concerns. So we must first look directly at problems with certain Liars in the light of it, before turning to a wider assessment of Read's paper.

3

The main question, of course, centres on what is wrong with the following kind of argument. From

   t = 't is not true',

there follows

   t is true ≡ 't is not true' is true.

But from the T-schema there follows

   't is not true' is true ≡ t is not true,

so it seems we can derive

   t is true ≡ t is not true.

There is no difficulty finding true identities like that expressed on the first line, and the second line follows necessarily from the first, the last from the preceding two, hence the problem must be with the third line itself. Read agrees, but what is causing the problem? As we shall see, it is the lack of a clear expression for the thought that t is not true which is keeping people gripped with the third line. One realizes there is a problem with getting a clear expression for such a thought not just by attending to the sort of grammatical difficulties in Read's paper discussed above. The difficulty is much more widespread, as can be seen if one attends even to the common description of sentences like t as 'self-referential'. One must remember, first of all, that the original identity does not itself show that some sentence is about itself, since it does not entail that

   $(\exists x)(x = $ 'x is not true'$)$,

by Existential Generalisation, and neither can 't' be replaced throughout by a quotation name for the supposedly self-referential sentence, since nothing of the form

   'p'= " 'p' is not true"

is possible: again, nothing can be a proper part of itself. By contrast, if one has a statement about *the content* of t, in a sentence where there is not *direct quotation*, but *indirect speech*, such as t says that t is not true,

there follows unproblematically that

$(\exists x)(x$ says that x is not true$)$,

and so that something is self-referential. And there is an equal possibility of providing a quotation name for a sentence of the required kind, since there is nothing against cases of the following form being true:

'p' says that 'p' is not true.

So properly there is no syntactic self-reference: no sentence in itself refers to itself.

That, indeed, may be granted quickly by many, saying that, of course, a sentence needs an interpretation before it can be said to be about anything, since it itself is just a syntactic string. But the rapidity of such an admission hides the crucial difficulty. The difficulty comes in various forms, and is at least threefold. First, in the non-professional area, people commonly do not attend closely to use-mention issues, so the syntactic string 't is not true' gets conflated with its *prima facie* interpretation, that t is not true, with the consequence that the expression " 't is not true' " gets used both for the uninterpreted string, and for the interpreted expression. Second, in the professional area of semantics, this conflation gets institutionalised when there is talk about sentences being 'true in models'. Is 'Pa' true if 'a' refers to Socrates and 'P' expresses the property of being wise? Why not say instead that what is true in the specified model is simply that Socrates is wise? The answer to that leads to the third, and major point enunciated before: in the professional area of formal logic there is considerable difficulty with saying the latter, since there is no standard expression, in the symbolism coming down from Frege, for the word 'that' in such expressions as 'that Socrates is wise'. Some notable individuals have followed Frege, who himself used a 'horizontal', and introduced a symbol for this purpose: thus Kneale used '§', and Cocchiarella 'λ' (Kneale 1972, Cocchiarella 1986, 217). But the generality of logic texts contain no such expression.

It is all this that combines to be the cause of the problem people have had with those Liar problems which involve direct quotation. There is another class of Liar problems involving indirect speech, which we shall look at later, but first we can clear up the problem with

't is not true' is true $\equiv$ t is not true.

For, because of the above, a question now arises about the referent of the " 't is not true' " on the left hand side. Either the quoted expression is a syntactic object, or it is an interpretation of that sentence which is meant, in which case the quoted expression is not a Tarskian syntactic object, but a Horwichian semantic one. In the Horwichian case, there is then an ambiguity, since in

t is true $\equiv$ 't is not true' is true,

the right hand side definitely involves a syntactic object, which means one cannot go on to get the traditional contradiction

t is true $\equiv$ t is not true,

by combining the two equivalences. The matter is even clearer if one removes the use-mention conflation in this case, and writes the T-schema equivalence as the Horwichian

that t is not true is true $\equiv$ t is not true.

Then it is very evident that there is no way to get the contradiction.

But in the Tarskian case, i.e. if it is a syntactic object in the T-schema, that again shows that we do not get a contradiction. For we would need in addition, as was pointed out before, a proof of the univocality of the sentence 't is not true', showing there is no other interpretation of it possible beyond the superficial one. Homophonic examples of the sentential T-schema are in the required way disquotational, but they cannot arise with semantically ambiguous sentences like 'There is a bank', or indexical ones like 'She is pretty'. So the question first has to be answered whether the same holds with 't is not true' in the special case where t='t is not true'. Once they are separated from their *prima facie* interpretations, however, it becomes clear that such sentences can have more than one interpretation. In the case of any sentence like 't is not true' what I have called above its *prima facie* interpretation, drawn from reading what is in its face, is that t is not true. But that depends on taking the 't' in it to refer to t, which, while certainly possible, is not necessary — as we saw before, there is no such thing as *syntactic* self-reference. Maybe at the top of some page there is the sentence 'the sentence at the top of the page is not true', but if a preceding sentence was, say, 'Once upon a time a wise old man could be found reading a certain page in a book', then the reference of the following subject term 'the sentence at the top of the page' is not the sentence at the top of the page that sentence is on, but a sentence in another possible world. And it is no good replacing that sentence with anything like 'the sentence at the top of the actual page this sentence is on is not true' since that involves an indexical, which could be given a variety of references, in different possible worlds. In these, and similar cases there is, or can be, therefore, a second semantic reading of the subject phrase, beyond the superficial one, showing no semblance of a paradox need be around.

Of course, there will be if the quoted 't' is given the same reference as the unquoted 't', but in that case, we must primarily remember that giving it that reference is a matter of choice. There is no syntactic self-reference, but also any semantic self-reference is not obligatory. Hence there is no way to get the T-schema absolutely, when the quoted expression on the left is a syntactic object, with the result that a contradiction does not necessarily follow this way, either. Isn't, at least, the sentence t, on the self-referential interpretation paradoxically both true and false? No, for what one is involved with, if one chooses the self-referential interpretation is not directly the syntactic identity, but the kind of statement about content which was introduced before, namely

t says that t is not true.

This is clearly true only on a certain semantic interpretation, and so it is not just an extensional remark about the syntactic object 't is not true'. The distinction makes it plain that what is true or false is not the sentence 't is not true', but the proposition that t is not true. As was pointed out before, the inability of the logical tradition to represent such a propositional referring phrase as 'that t is not true' has made it seem that what is true or false on the given interpretation is still the (mentioned) sentence, but only the sentence in use, preceded by 'that', refers to the item that has the truth value. In fact, as a result, sentences, in themselves, are neither true nor false, so that t is not true is definitely true, in this case.

It follows that there is no longer any problem with Strengthened Liars. Certainly the question naturally arises about what to say in 'strengthened' cases, where, for instance,

s = 's is neither true nor false'.

But first of all there is no need to take the sentence thus defined to speak about it itself being neither true nor false. For sentences, by themselves, have no voice. If anyone chooses to interpret this sentence self-referentially that is therefore an additional (intensional) matter beyond the direct speech identity. And what is true in that self-referential case is simply that s is neither true nor false, which is not paradoxical in any way, since to say that is not to say that what is true is s, i.e. 's is neither true nor false'. The self-referential case, moreover, is expressed with

s says that s is neither true nor false,

so it is to a study of such expressions in indirect speech, which form a large part of Read's paper, that we now turn.

4

The above supports very forcibly Read's arguments against the Tarskian T-schema, but, as it stands, it leads to a positive account of truth that, while very similar to Read's, is crucially different from it. Read's case (C), for instance, is one of those covered, in a very similar way, in Goodstein's original paper on the formalisation of indirect speech (Goodstein 1958), which led directly to Prior's theory (Prior 1971), and then eventually to its subsequent modification using Kneale's 'that'. But, for one thing, there are some grammatical problems with Read's symbolisation of this same kind of material. More importantly, he also introduces, or at least relies on, a further axiom that rules out the kind of ambiguity we have seen to be necessary to allow an escape from the above syntactic paradoxes.

What are the grammatical difficulties when we move over to Liar problems arising with the indirect speech form 'x says that p', if this is symbolised as Read symbolises it: 'x:p'? For a start, the previous difficulties with formalising 'that p is true' affect this style of symbolisation, in formulas such as Read's (A), i.e.

$Tx \equiv (\forall p)(x{:}p \to p)$.

This expression employs quantification over propositions, which Read glosses with a reference to Church's type theory, in footnote 2. This kind of symbolism was used at length by Prior, and indeed Read's thesis (A) is to be found in Prior's book *Objects of Thought*, on page 104. But there are major problems with reading Prior's propositional quantifiers, which he himself struggled with in chapter 3 of his book (see also Haack 1978, 130), and these are also present in Read's account. Centrally, there is no provision, within such a system of propositional quantification, for truth predications on propositions. In 'x:p & p', for example, the 'p' is not a referential phrase, and so any 'p' in a quantifier before it equally would have to be not a referential phrase; also when reading 'x:p & p' as 'x says that p and it is true that p', there is no representation of the 'it is true that' in the formalism.

What is wanted, first of all (c.f. Slater 2001), is a nominaliser, which will produce, in a formalisation of 'that p' (here 'λp'), a referring phrase to a proposition expressed by 'p'. In addition one then needs a truth predicate of such nominalised phrases, and, when generalising expressions like 'x says that p, and that p is true', one must employ variables which range not only over 'that' clauses, but also other propositional referring phrases, like 'what x says', for instance. The form of the given case is then 'x says r, and r is true' over which one can quantify objectivally, with quantifiers which read quite straightforwardly 'for some/for all propositions r', etc. The analysis of Read's (C), i.e.

$$\forall p(C{:}p \rightarrow \neg p),$$

then proceeds as follows. First C says that everything that C says is false, i.e.

$$Sc\lambda(\forall r)(Scr \supset \neg Tr),$$

so suppose that everything C says was indeed false, i.e.

$$(\forall r)(Scr \supset \neg Tr),$$

then, contradictorily, something C says would be true, i.e.

$$(\exists r)(Scr \,\&\, Tr),$$

since from the propositional T-schema there follows in particular that

$$(\forall r)(Scr \supset \neg Tr) \equiv T\lambda(\forall r)(Scr \supset \neg Tr).$$

Hence not everything C says can be false, i.e.

$$\neg(\forall r)(Scr \supset \neg Tr),$$

and so, equivalently,

$$(\exists r)(Scr \,\&\, Tr).$$

But then something C says is false, namely

$$\lambda(\forall r)(Scr \supset \neg Tr),$$

and so, also,

$$(\exists r)(Scr \,\&\, \neg Tr).$$

It follows that C must say (indirectly) at least two distinct things, one true and the other false. Read says similarly 'However, although C:$\forall p(C{:}p \rightarrow \neg p)$ this may not be all that C says' obtaining, as well, $(\exists p)(C{:}p \,\&\, p)$.

Indeed, there is a close similarity with another matter raised at the start, since from Read's (A), i.e.

$$Tx \equiv \forall p(x{:}p \rightarrow p),$$

there follows a conditionalised T-schema, requiring 'singleness of saying', i.e. something like univocality, or uniqueness of interpretation, before there can be truth assessments of sentences in the traditional T-schema form, viz:

$$[x{:}p \,\&\, (q)(x{:}q \rightarrow q{=}p)] \supset (Tx \equiv p).$$

So both the necessity of this conditionalised T-schema, and its consequences, are endorsed by Read, and they are to be accepted, once grammatical corrections have been made.

What is there to be said, by contrast, in substantive criticism of Read's treatment of Liar paradoxes arising in indirect speech? The main point which needs to be made relates to this issue of univocality, since the contrary line of analysis allows for ambiguous sentences, and even requires them in certain places, while Read explicitly dismisses ambiguity, and holds that if x both says that p and says that q, then x says that p and q. Certainly this latter rule, namely Adjunction, is not made explicit in Read's axioms, but he still employs it, making, for instance, his transposition, in his section 4, from

L:(TL v ¬q),

to

L:((TL v ¬q) & q),

given L:q, not something actually enforced by his stated rules. Read uses Adjunction, implicitly, in this derivation in his section 4, and in his discussion of (C) in his section 5. Thus, after the remark about C:∀p(C:p → ¬p) possibly not being all that C says, quoted above, he goes on: 'Suppose C also says that q, that is C:(∀p(C:p → ¬p) & q)', adding the extra 'q' as a conjunct to the other thing C says. But, for a start, given that so much can be obtained without the use of this rule (see e.g. Slater 2004a), Ockham's Razor, if nothing else, suggests strongly that it should be dispensed with. One is reminded, on that score, of the discussion of Moore's Paradox in the 'Budget of Paradoxes' chapter of Prior 1971, 81-84. This is first analysed using Hintikka's axioms for belief, but then an entirely adequate explanation of its 'logical oddity' is produced using nothing more than quantification theory.

There is a stronger argument against Adjunction, however, than the need for axiomatic economy. For we have seen, when dealing with Liar paradoxes arising in direct speech, that it is necessary to grant ambiguity a place, in connection with identities like

t = 't is not true'.

So why is Read against ambiguity? Read does debate the matter, but only briefly, and his conclusions are not well supported by his arguments. Thus he is concerned that the point of formalisation is to remove ambiguity, but seems to forget that that concern still allows entirely precise statements to be made about the ambiguity of certain sentences. If x is ambiguous, it can still be completely unambiguous that x has two meanings, i.e. means that p and means that q without meaning that p and q. More specifically, Read says, about ambiguous expressions such as 'Visiting relatives can be boring', that 'we do not require that both senses be the case for x to be true — either will suffice'. But such a sentence is then true in one sense and false in another, not true simpliciter either time, making Read's 'Tx' better read 'x is true in all senses', not 'true simpliciter'. Only starting from the conditionalised truth schema identified above is the truth of sentences limited to truth simpliciter, and so to truth as more generally understood within the Tarskian tradition. And there is another notion that is more generally part of that tradition. For the discussion above, for instance, involved the notion of meaning, which is a notion not found in Read's paper, but crucial to the Davidsonian tradition following on from Tarski. Read glosses his 'x:p' in various ways, such as 'x says that p', 'x implies that p', but there is another expression in the area, 'x means that p', which behaves rather differently. Thus 'John is a bachelor' means that John is an unmarried male of marriageable age, and the latter implies that John is unmarried. So the meaning of something is its entire content, or, at least, something from which its entire content can be deduced. Writing 'x means that p' as 'Mxλp', and using 'I' for the relation of implication, we can in fact then define 'x says that p' as 'something x means implies that p', i.e.

$(\exists r)(Mxr \ \& \ Ir\lambda p)$.

Given this, the other main axiom Read employs, (K), i.e.

$(\forall p,q)((p \supset q) \supset (x{:}p \supset x{:}q))$,

becomes, with appropriate grammatical corrections, just a matter of definition also. Thus we can get 'x says whatever is implied by what it says', as the schema:

$(p \supset q) \supset [(\exists r)(Mxr \ \& \ Ir\lambda p) \supset (\exists r)(Mxr \ \& \ Ir\lambda q)]$.

This follows from the transitivity of implication, using some quantification theory, and attending to the appropriate grammar. It must be remembered, in this connection, that 'implies' is a verb and relates two propositional subjects given by referring phrases, as in 'that John is a bachelor implies that he is unmarried' and 'what Peter stated implies that John is unmarried'. The symbol which Read uses, '$\rightarrow$', is, by contrast, a propositional connective, which can be defined via the equivalence:

$I\lambda p\lambda q$ if and only if $p \rightarrow q$.

Lack of a propositional nominaliser, of course, is what inclines many (indeed most) to read the connective as 'implies'. That confusion only deepens if one says, like Read, that sentences imply things; indeed one is then involved in the full grammatical confusion with which we started.

5

In conclusion, I have supported Read's arguments against Tarski, pinpointing more accurately the fallacy that holds people to his schema a certain conflation of syntax and semantics. However, the core grammatical insight needed to be clear of Tarski is still widely ignored in Read's system, and there are superfluous and restrictive axioms there that either can be reduced to definitions, or do not allow for the full extent of the cases that can be encountered. Logic is an exact science, but there are many problems with being as exact as it requires, and so Logic must allow for inexact and ambiguous expressions. Only quantification theory, at the propositional, i.e. indirect speech level, is required to work through the 'paradoxical' cases that result. Read's rejection of ambiguity and his consequent endorsement of Adjunction, are the central features of his account which prevent him from seeing that only quantification theory is needed. But his lack of appreciation of the place of ambiguity is also directly linked to his confusion over the sentential versus propositional nature of the T-schema. And we have seen, at length, that a clarification of that distinction is what gets one out of standard Liar paradoxes in direct speech.

CHAPTER 6

TRANSLATABLE SELF-REFERENCE

1

Stephen Read has advanced a solution of certain semantic paradoxes recently, based on the work of Thomas Bradwardine. One consequence of this approach, however, is that if Socrates utters only 'Socrates utters a falsehood' (a), while Plato says 'Socrates utters a falsehood' (b), then, for Bradwardine two different propositions are involved on account of (a) being self-referential, while (b) is not. Problems with this consequence are first discussed before a closely related analysis is provided that escapes it. Moreover, this alternative analysis merely relies on quantification theory at the propositional level, so there is very little to question about it. The analysis is the third in a series explaining the superior virtues of a referential form of propositional quantification.

2

Stephen Read has advanced a solution of certain semantic paradoxes recently, based on the work of Thomas Bradwardine (see, for example, Read 2002, 2008a, 2008b, 2010). Read claims that Tarski's T-scheme, in the form (T) $T(s) \equiv A$, must be improved so that, while Tarski "said that (T) should hold when what replaces 'A' is a translation (into the metalanguage) of the sentence whose name is 's' Bradwardine's point is that what replaces 'A' in (T) must express everything that the utterance named by 's' expresses or signifies " (Read 2010, 166). Writing 'Sig(s, A)' to express the fact that sentence ('utterance-type') s signifies that A, Read defines the truth and falsity of sentences in the following fashion:

$Tr(s) \equiv (\exists A)Sig(s, A)$ & $(\forall A)(Sig(s, A) \supset A)$,

$Fa(s) \equiv (\exists A)(Sig(s, A)$ & $\neg A)$.

He then takes Bradwardine's major principle to mean that if A entails B then Sig(s, A) entails Sig(s, B), so that while, for instance, in a paradoxical case, Tarski would think that a given sentence meant just that it was false, or not true, Bradwardine would start from the assumption that the sentence might express or signify not only that but something else, as well: Sig(s, Fa(s) & Q). In fact it follows from Bradwardine's principles that no sentence can signify just that it is false, or not true. After resolving an impressive number of puzzles on this basis, Read has summed up: 'Once truth and meaning are properly understood, so that an utterance is true only if everything it signifies obtains, paradox is prevented without need for logical revision' (Read 2010, 175-6).

One consequence of this approach, however, is that if Socrates utters only 'Socrates utters a falsehood' (a), while Plato says 'Socrates utters a falsehood' (b), then, for Bradwardine two different propositions are involved on account of (a) being self-referential, while (b) is not. It cannot be the same proposition that is involved since, on Bradwardine's account, the first is false while the second is true. Thus Read says (Read 2010, 171 though take care, the two truth-values of (a) and (b) are accidentally reversed on line 3 of this page):

> Bradwardine's response was that (a) and (b) were not the same proposition, since (a) is self-referential, saying of itself that it is false, and so, by Thesis (2), saying also that it is true. That is why it is false. But (b) is not self-referential, but says (only) that (a) is false (and uttered by Socrates). Superficially, (a) and (b) are tokens of the same utterance type, so it might seem that we have here a counterexample to my earlier claim that Bradwardine attributed truth and signification to utterance types. However, we have just seen that (a) and (b) differ in signification. So they are tokens of different sentence-types.

But what is another token (if there is one) of the sentence-type that Socrates is then supposed to utter? Since (a) is a token sentence, Socrates' uttering the same words again would, on Read's analysis, involve self-reference to another, different token, and so could not have the same signification as (a). What does have the same signification, supposedly, is '(a) is false and (a) is true and (a) is uttered by Socrates'. But that is certainly not of the same sentence-type as (a).

And does it even have the same content? That is the more important question. Read seems, in fact, to be forgetting the rules about translation into reported speech, and specifically confusing a proper name with an egocentric particular. One must take the utterer into account to get the content of what is said if the speaker says 'I am uttering a falsehood'. But nothing needs to be changed if a proper name is involved, as in 'Socrates utters a falsehood'. How else is anyone going to report *what Socrates said* in the given case if not by uttering what Plato did, or its equivalent?

Maybe Read, after a re-consideration, would like to say that what Socrates says in the case above is in fact untranslatable. This option is taken by 'cassationists' (see, for instance, Goldstein 2006), and would mean saying that Socrates' utterance did not express a proposition, making Plato's remark false, since Socrates' was neither true nor false. But surely Socrates agrees with Plato; he can even say 'Yes!' or 'That's true' after Plato's utterance. So Socrates must be able to have the same thought as Plato; indeed, as before, the rules of reported speech require that in the context above both have the same thought. So it has to be pointed out that there is another very similar treatment of the same puzzles Read has been dealing with that does not get into the same difficulties (see, for example, Slater 2001, and chapter 4).

I use an improved form of Prior's propositional quantification incorporating Coccharella's nominalising functor 'λ' ('that'), which makes 'λp' a phrase referring to the proposition expressed by 'p' on the occasion of its use, with the standard interpretation. Propositional identity statements of the form '$\lambda p = \lambda q$' are true only if 'p' and 'q' are synonymous, which involves a tighter relation than logical equivalence. But there are many other propositional identity statements that are not of this form, since there are many other ways to refer to propositions than by using 'that'-clauses. Here I also use the epsilon calculus to provide other propositional referential terms; together these two improvements enable one to generate all the relevant propositional identity statements that are needed. As we shall see, it is the provision of propositional referential terms that chiefly separates this alternative treatment from Read's, and specifically in the above and similar cases it allows the extra significance of the puzzling utterances to be pointed towards while leaving it un-expressed, thereby capturing in a more appropriate way its 'non-translatability'. I write 'Sxr' for '(sentence) x signifies (proposition) r', so that 'ϵrSxr' means 'what x signifies' (which might be a matter of choice if there are several things x signifies). General equivalences are then that $T\lambda p \equiv p$, and that $Sx\epsilon rSxr \equiv (\exists r)Sxr$.

Read does not use propositional referential terms, as we saw at the start; nor does he have a truth predicate with propositions. Thus his analysis involves quantification over expressed sentences, as with '$(\exists A)Sig(s, A)$', in the manner of Prior. Also the truth of propositions is not represented as a predicate, but simply by the assertion of the proposition, if at all. That leads him to bring in what extra might be involved in a case of self-reference by means of some further expressed sentence: viz '$Sig(s, Fa(s) \& Q)$'. From this, using the principles indicated above, Read easily draws the conclusion $Fa(s)$ (Read 2010, 169), leading him to take this to be the case when Socrates utters 'Socrates utters a falsehood', but, of course, not to be the case when Plato utters the same thing. So he is drawn to analyse Socrates' remark as though it was untranslatable, in the manner we saw before.

A couple of elementary results, for a start, will show how the closely related approach using propositional referential terms does not have this unfortunate consequence. First it is true that if a sentence x signifies that what that sentence signifies is false (or not true) then there is not just one thing the sentence signifies, but one thing that is true and another thing that is false. This would arise, for example, with 'What this sentence signifies is not true', on a self-referential reading of 'this sentence'. For if

$$Sx\lambda(\exists r)[Sxr \& (t)(Sxt \supset t=r) \& \neg Tr],$$

then, if

$$(r)(Sxr \supset Tr),$$

then, by substitution,

$$T\lambda(\exists r)[Sxr \& (t)(Sxt \supset t=r) \& \neg Tr],$$

and so in particular

$$(\exists r)(Sxr \& \neg Tr),$$

a contradiction. So

$$\neg(r)(Sxr \supset Tr),$$

i.e.

$$(\exists r)(Sxr \& \neg Tr).$$

But the latter, together with

$$Sx\lambda(\exists r)[Sxr \& \neg Tr],$$

means that

$$(\exists r)(Sxr \& Tr).$$

But that means that
    (∃t)(∃r)(Sxt & Sxr & ¬Tt & Tr),
and so
    (∃t)(∃r)(Sxt & Sxr & t ≠ r).
Of course from
    (∃r)(Sxr & ¬Tr)
one can get
    Sxεr(Sxr & ¬Tr) & ¬Tεr(Sxr & ¬Tr),
while from
    (∃r)(Sxr & Tr)
one can get
    Sxεr(Sxr & Tr) & Tεr(Sxr & Tr),
so identifying leading candidates for εr(Sxr & ¬Tr) and εr(Sxr & Tr) is quite
straightforward since
    Sxλ(∃r)[Sxr & (t)(Sxt ⊃ t=r)],
while
    ¬Tλ(∃r)[Sxr & (t)(Sxt ⊃ t=r)],
and
    Sxλ(∃r)[Sxr & ¬Tr],
while
    Tλ(∃r)[Sxr & ¬Tr].
Notice also that parallel things would arise given
    Syλ(∃r)[Sxr & (t)(Sxt ⊃ t=r) & ¬Tr],
i.e. with, for example, 'What that sentence signifies is not true' with the
previous sentence as the referent of 'that sentence'. For
    Syλ(∃r)[Sxr & (t)(Sxt ⊃ t=r)],
and
    Syλ(∃r)[Sxr & ¬Tr],
and what y says here are, respectively, false and true. So in this case the report
given by y carries completely the same message as the original expression by
x, i.e. the content of x is unproblematically translatable.

    A different result is obtainable supposing merely that a sentence signifies
that something it signifies is not true, i.e.
    Sxλ(∃r)(Sxr & ¬Tr),
which is the same case as that above without the uniqueness clause. Certainly
then, again, if
    (r)(Sxr ⊃ Tr),
then, by substitution,
    Tλ(∃r)(Sxr & ¬Tr),
and so
    (∃r)(Sxr & ¬Tr),
a contradiction. Hence
    ¬(r)(Sxr ⊃ Tr),
i.e.
    (∃r)(Sxr & ¬Tr).
And the latter, together with what was explicitly said in this case, means that
    (∃r)(Sxr & Tr).

So again something x signifies is false, and something x signifies is true. But while, again, one thing that is said that is true is the content of the sentence, i.e.

$$\lambda(\exists r)(Sxr \ \& \ \neg Tr),$$

whatever is said that is false is not immediately expressible in a 'that'-clause, but instead is just referred to by the epsilon term in

'$Sx\epsilon r(Sxr \ \& \ \neg Tr) \ \& \ \neg T\epsilon r(Sxr \ \& \ \neg Tr)$'.

And now no entirely comparable result is entailed if

$$Sy\lambda(\exists r)(Sxr \ \& \ \neg Tr),$$

since then, while it follows that

$$(\exists r)(Syr \ \& \ Tr),$$

there is no way to obtain that

$$(\exists r)(Syr \ \& \ \neg Tr).$$

Turning now to Read's case above, we find it is the same as this last one, since it is about a sentence that says not about what it says, but about itself that it is false. Writing

$$Vs \equiv (r)(Ssr \supset Tr),$$

then given $Ss\lambda\neg Vs$, we immediately get from the last case that

$$(\exists r)(Ssr \ \& \ \neg Tr),$$

and that

$$(\exists r)(Ssr \ \& \ Tr).$$

The former result shows that in fact $\neg Vs$ (Read's '$Fa(s)$'), and it is from this fact, with $Ss\lambda\neg Vs$, that we get that there is something Socrates says that is true, namely

$$\lambda(\exists r)(Ssr \ \& \ \neg Tr).$$

Given $Sp\lambda\neg Vs$, that means also that

$$(\exists r)(Spr \ \& \ Tr),$$

for the same substitution instance, showing Socrates and Plato do then say one and the same thing that is true. But what it is that Socrates implies that is false in uttering what he did is not so easily locatable, and in fact what anyone might add to get something equivalent to the whole of what Socrates is implying is not obtainable from the given linguistic context, since it is not expressed explicitly there.

It is at this point that the present account wins out over Read's, since there is still something Plato might add to express an equivalent to what Socrates is thinking — if only Socrates would make it explicit in a 'that'-clause. So what Socrates is thinking as a whole is not untranslatable; it is merely not wholly translatable given the limited description of the context presented above. The central formal difference from Read's account which allows for this is that the limited context above, even if it does not give us what proposition is implied that is false, still does give us *a means of referring to that proposition,* and also ensures us that that proposition is a different proposition from the one explicitly made. For

$$\neg T\epsilon r(Ssr \ \& \ \neg Tr),$$

while

$$T\lambda(\exists r)(Ssr \ \& \ \neg Tr).$$

But only the speaker, Socrates, can help us to locate $\epsilon r(Ssr \ \& \ \neg Tr)$ explicitly, and so express it, since by just uttering 'Socrates utters a falsehood' he has yet to spell out which falsehood is on his mind. In this connection it is important to note that while the remark is 'self-referential' in that it refers to the speaker making it, and also to the sentence that is used to express it, there is no indication that what proposition is taken to be false is what is then said, i.e. that

$Ss\lambda(y)(Ssy \supset y = \lambda(\exists r)(Ssr \ \& \ \neg Tr))$.

If that had been added then we would have a case like the initial one analysed, where there was an explicit mention of something said that was false. But the case Read has described is strictly without this further form of 'self-reference', making what is implied that is false indeterminate, from what has so far been given.

Quite generally, if a speaker makes an existential remark like '$(\exists r)Pr$', then we can ask 'Which P is it?' with the 'it' then in question being tracked by the associated epsilon term 'ϵrPr'; and if a speaker says '$\neg(r)Pr$' then we may ask 'Which is it that is not P?', with the 'it' being tracked by '$\epsilon r\neg Pr$'. It is something of the former kind which is a candidate for being the extra, unexpressed thought Socrates has that is not true when he utters '$(\exists r)(Ssr \ \& \ \neg Tr)$'. *That Socrates was thinking something that is false* is demonstrably true, but he needs to tell us emphwhat he was thinking that he was thinking and was false, before we have a chance of telling him explicitly what he was thinking that is false. There are two cases, the first being the case above where uniqueness is a part of the very thought he is then actually having, and is just what is false. That is to say we have

$Ss\lambda(Sst \ \& \ \neg Tt)$

with

$t=\lambda(\exists x)(Sst \ \& \ \neg Tt)$

while

$T\lambda(\exists x)(Sst \ \& \ \neg Tt)$.

But it also may be that he thinks he is saying something else by the remark we have been considering, and thinks that *that* proposition is false, i.e. that

$Ss\lambda(Sst \ \& \ \neg Tt)$

for a quite separate

$t\neq\lambda(\exists x)(Sst \ \& \ \neg Tt)$.

Since, however, he is not saying anything else, this saying is imaginary and we have

$\neg(Sst \ \& \ \neg Tt)$,

so

'$\lambda(Sst \ \& \ \neg Tt)$'

provides the needed truth maker for

'$(\exists t)(Ssr \ \& \ \neg Tr)$'.

One needs to be familiar with the epsilon calculus to realise fully that making an existential remark always means referring, or at least alluding to an instance, since the predicate calculus hasn't got an automatic placeholder for what individual thing is being spoken about when someone makes such a remark. Intuitionists do not appreciate this point fully, and in particular would have

difficulty with a request for a witness to a negated universal remark in the manner presented before. The epsilon calculus does not sit easily with Intuitionist
Logic (see Slater 2009b, 397-402 for a full discussion) and the matter is settled
against this logic if it is granted that the previous question 'Which is it that is
not P?' can be appropriately asked after a negated universal remark. For even
if the identity of $\epsilon r \neg Pr$ is unknown, at least it is then granted that $\neg P\epsilon r \neg Pr$,
and so $(\exists r)\neg Pr$. But the further problem in Read's case is that the existential
quantifier is within the scope of an intensional operator, so any instantiation
which the speaker has in mind is up to him to say, and remains a feature of
the broader pragmatic context. It may seem that there is nothing similar in
the earlier case without a uniqueness clause, since no speaker was nominated,
and how could a bare sentence provide an explicit answer to what further it
implicitly means in that case? But no sentence on its own is self-referential (see
Slater 2002a), since one cannot, for instance, derive by objectival Existential
Generalisation "$(\exists z)(z='(\exists r)(Szr \ \& \ \neg Tr)')$" from "$x='(\exists r)(Sxr \ \& \ \neg Tr)'$ ". The
presence of the quotes prevents this, but also, thereby the 'x' in quotes might
refer to a quite different sentence, as, equally, 'this sentence' in 'what this sentence signifies is not true' would do if it were accompanied with a gesture to
another sentence. So to get the supposed referent of 'x' and make it the case
(in indirect speech) that

$Sx\lambda(\exists r)(Sxr \ \& \ \neg Tr)$,

the sentence

'$(\exists r)(Sxr \ \& \ \neg Tr)$'

must be uttered by someone giving it a self-referential interpretation. Therefore
to get an answer to the question about what $\epsilon r(Sxr \ \& \ \neg Tr)$ is, one must ask
that speaker.

The need to consider the speaker's intentions to secure the self-reference is
centrally relevant in connection with other cases, for instance when speaker M
says 'Everything thought by M is false'. For in order to ensure that the quoted
'M' is used to refer to the speaker in question the speaker must realise that the
universal thought expressed here is one of his or her thoughts. Forgetting this
we might symbolise the content of the expression by just reporting:

$Sm\lambda(r)(Smr \supset \neg Tr)$.

The analysis of this proceeds much as before, since if we suppose that

$(r)(Smr \supset \neg Tr)$

then, by substitution

$(\exists r)(Smr \ \& \ Tr)$,

a contradiction, proving that

$(\exists r)(Smr \ \& \ Tr)$.

But that, together with the initial report then means that

$(\exists r)(Smr \ \& \ \neg Tr)$.

The instance that verifies the latter existence claim is thereby assured, but
what is the instance that verifies the former existence claim, i.e. that

$(\exists r)(Smr \ \& \ Tr)$?

It is not available from the context as so far formalised, and has to be derived
from what the above incomplete report has missed out:

$Sm\lambda Sm\lambda(r)(Smr \supset \neg Tr)$.

For that, with what was previously reported entails

(∃r)(Smr & Tr).

A similar implication was available in the other case, namely that

SsλSsλ(∃r)(Ssr & ¬Tr),

although this was not needed in the analysis above.

To think that some sentences involved in problems to do with self-reference must have a speaker whose unvoiced thoughts are relevant to the case is not commonly realised. But I suppose some people might also have to adjust to the complex, delicate subtlety of the analysis of this sort of situation that is now clearly available formally. It really is quite remarkable, I agree. But it is just a symbolisation of natural language, so it is that natural language that such people should better respect. For the record, many other paradoxes are dissolved this way, such as Curry's Paradox (which Read deals with in Read 2010) and also, for instance, The Knower Paradox, and Gödel's Paradox, in Slater 2004a. Two other major problems for Priorian propositional quantification I have shown to disappear with the more sophisticated, referential version, in Slater 2001, and chapter 4 — Read has yet to address either of these issues. The only relevant paradox I have not previously discussed is Field's Paradox, but that is resolved in much the same way as in Read 2010, 175.

CHAPTER 7

WHAT PRIEST HAS BEEN MISSING

1

It is shown that there are categorical differences between sentences and statements, which have the consequence in particular that there are no paradoxical
cases of self-reference with the latter as there are with the former. The point
corrects an extensive train of thought that Graham Priest has pursued over
recent years, but also a much wider tradition in logic and the foundations of
mathematics that has been dominant for over a century. That tradition might
be broadly characterized as Formalist, or Nominalist, and the improved understanding of statements leads us instead into a more Realist approach and
thereby contentful logic and mathematics.

2

Other people besides Priest have been involved in the following matters, but
Priest's treatment has been most extensive, extreme, and persistent, so it will
be useful to pursue the general discussion in connection with just his work.
And what has held Priest to his extraordinary beliefs is primarily an error far
more widely committed.

One of Priest's latest arguments in defence of his 'Dialetheism', for instance,
is his claim that Boolean negation (i.e. what has traditionally been called
'contradiction') has no sense. This claim is to be found in *Doubt Truth to be
a Liar* where, after showing how the introduction and elimination rules for
Prior's connective 'tonk' deliver triviality, Priest tries to show 'The same, in
the appropriate context, is true of Boolean negation' (Priest 2006, 94). The
context Priest thinks appropriate consists in harmonious introduction and elimination rules for Boolean negation, together with those for Truth, and a case of
paradoxical self-reference. Others might derive from this combination a need
to modify the given rules for Truth, but Priest thinks the resulting triviality
shows instead that Boolean negation is incoherent. Rather, as we shall shortly
see, it is his case of paradoxical self-reference which has to be questioned.

What Priest, amongst many others, has been missing is evident earlier in
the same book, where he says (Priest 2006, 43).

> Characterising contradiction is relatively easy: contradictories are
> any things of the form α and $\neg\alpha$. This definition hides a difficulty,
> though. What sort of things are we talking about here: sentences,
> statements, propositions, beliefs? This is a thorny issue. Fortu-
> nately, then, nothing much seems to turn on the niceties of the
> question for present purposes. I will simply assume that α and its
> ilk are truth-bearers, whatever these are required to be.

51

Here Priest missed Feferman's short proof, and also Koslow's longer proof, (Feferman 1984, 245, Koslow 1972, 291) that there is no Fixed Point Theorem for the truth operator or at least the relevance of this to truth as a property of propositions or statements. We shall look at the Fixed Point Theorem in more detail later, but the essential point is that there are categorical differences between sentences, statements and propositions, as Priest's one-time mentor Susan Haack, following Strawson and Lemmon, explained, and as we shall now see even more clearly (Haack 1978, Ch 6, Lemmon 1966). Haack used a symbolism derived from William Kneale, below I shall use one more recently popularized by Paul Horwich (Kneale 1972, Horwich 1998). Thus Horwich's '<p>' means 'the proposition that p', or simply 'that p'. Some of the above thinkers prefer to restrict the term 'proposition' to Strawson's notion of the general sense of a sentence (comparable to Kaplan's 'character' (Kaplan 1989)), with 'statement' then being the word for what has the truth-value (i.e. being comparable to Kaplan's 'content'). That will emerge as crucial at one point later, but for a start the two terms will be used more interchangeably.

For what is more important first of all is to remember the distinctive way we talk explicitly about sentences: by means of referring phrases like 'snow is white'. By contrast, to refer explicitly to a proposition or statement we use forms like 'that snow is white'. The difference is thus marked by the difference between direct and indirect speech, and, as we shall see, it is only with direct speech forms that we get paradoxical self-reference. But even that is removed once one attends to the propositions or statements then made.

For instance, a paradox arises if the sentence at the top of the page is 'the sentence at the top of the page is not true'. In more general terms, we can easily construct a sentence J such that J='J is not true'. And Tarski's Truth Scheme (see Tarski 1956) if applied within the same language, says that 'p' is true if and only if p, which entails that 'p' is not true if and only if it is not true that p. So that leads to the well-known Liar Paradox via the series of equivalences: J is not true if and only if 'J is not true' is not true, i.e. if and only if it is not true that J is not true, and so if and only if J is true. We shall see later just why Tarski's Truth Scheme does not apply, but suppose we use Horwich's Equivalence Scheme instead: <p> is true if and only if p. Then, obviously, we do not get a contradiction. We can say that sentence J is not true, since now sentences do not have any truth-value. But since sentence J is not true if and only if the sentence 'J is not true' is not true, and not if and only if the proposition <J is not true> is not true, we cannot deduce that J is true. So the contradiction Tarski derived on the supposition of semantic closure within the same language arose through a category mistake, and specifically an use-mention confusion. If the sentence 'J is not true' was the proposition <J is not true>, then there would be a contradiction, but that would involve equating a mentioned sentence with a 'that'-clause.

3

There is a further, much more general argument, however, showing that no contradictions need arise on a Horwichian account of truth. That comes from noting the following piece of grammar. For <p> is true if and only if that p is true, i.e. if and only if it is true that p, while 'It is true that p' would standardly be written in modal fashion: 'Lp'. That is to say, Horwich's 'That p is true', while it is of the subject-predicate form, is equivalent to 'It is true that p', which is of an operator form. This is a quite general relation (see the O.E.D., for instance, under 'that' as a conjunction) linking non-cleft forms with cleft forms starting with a waiting pronoun; thus 'He is wealthy, my uncle' is equivalent to 'My uncle is wealthy' on the same principle. But 'It is true that p' involves the operator 'it is true that' which is the null or identity operator in the modal system KT. And one modal fact in KT is that if <Lp> is equivalent to <p> then it is not the case that <p> is equivalent to <not-Lp>, since the modal system KT is consistent. So <p> can never be equivalent to <<p> is not true>, and the conclusion must be that there are no paradoxical, self-referential propositions (and indeed that there are no 'fixed points' for propositions).

But how can it be that there are no paradoxical self-referential propositions? We can easily form a sentence J* such that J*='the statement made by J* is not true'. Horwich has considered another case of the form 'THE PROPOSITION FORMULATED IN CAPITAL LETTERS IS NOT TRUE' (Horwich 1998, 41-2). Haack's puzzling case concerns a sentence numbered 1='the statement made by the sentence numbered 1 is not true' (Haack 1978, 150). Surely there must be a contradiction with these? Don't we get, each time, that that there is some proposition or statement G such that G=<G is not true>, and therefore that G is not true if and only if <G is not true> is not true, and so if and only if G is true? Anyone who thinks so, however, is forgetting the possibility that no definite statement is made in such cases, as there would be if the referring phrase ostensibly referring to a proposition or statement were replaced by a non-descriptive demonstrative such as 'this'. For then it would follow that the sentence on its own outside some pragmatic use would not state anything definite to be not true: 'This is not true', while it has a general sense, does not itself make a statement, without a referent for 'this' being given. It follows by *Reductio* from the supposed contradiction above, therefore, that in the given place the definite description 'the statement made by J*' must function like 'this', and so be non-attributive in Mill's and Donnellan's sense (Donnellan 1966). That is to say, it must be such that its reference is not determinable from its sense, and specifically such that its reference need not be any single statement made by J*. Likewise with Horwich's and Haack's referential phrases.

The point applies more broadly, of course. Thus suppose there was an enumeration of sentences in which sentence 2 was 'the statement made by sentence 2 is true'. Then if the statement made by sentence 2 were that the statement made by sentence 2 is true, the supposed statement would not be paradoxical by having to be equivalent to its contradictory. But still we would have a circular definition of the supposed statement leading to an infinite regress. So the central point in all such cases is that there cannot be a determinate statement in connection with them. But how can this be? That is the main puzzle.

Clearly the circularity arises because an indirect speech construction is involved, unlike when the sentence at the top of the page is 'the sentence at the top of the page is not true'. But still it may seem that a definite statement is made, and it is here where the proposition-statement distinction must be insisted upon. For certainly the sentence has a general sense, but that does not mean that a definite truth-value can be associated with it. So what must be recognised is that referential phrases like 'the statement made by the sentence numbered 1', 'the statement made by sentence 2' share the indexical nature of demonstratives like 'this', at least to the extent that any referents they are given are only determinable from some pragmatic use, where the intended subject is pointed out *in the wider context.* Only by remembering Mill and Donnellan does it become entirely clear why referential terms quite generally do this. For while the sentence 'the statement made by sentence 2 is true' contains the words 'the statement made by sentence 2', that does not mean there is any one such statement, and, in fact, different statements, or no statement at all, may be made with *the very same sentence* 'the statement made by sentence 2 is true' dependent on what the referential phrase 'sentence 2' in it is taken to refer to, i.e. on how, *in the wider context,* the supposed enumeration of sentences is arranged. This kind of indexicality is why Tarski's Truth Scheme did not apply to 'the sentence at the top of the page is not true' and 'J is not true' because 'the sentence at the top of the page' can be used with respect to the top of many pages, and 'J' with respect to many sentences, and even more determinate 'eternal' forms like 'the sentence at the top of page n of book B' are readily seen to be indexical, in the same sense, once we remember they may refer to different sentences in different possible worlds.

The overriding general principle is that propositions and statements, unlike sentences or their tokens, are not empirical objects: they are intentional objects that obey only the laws of logic. So before they can be identified a proof of their existence must first be obtained, i.e. it must first be shown that some consistent sense (and just one consistent sense) can be made of the associated sentences, in the given context. Haack in particular was confused on this point, since she claimed that the statement made by her 'sentence numbered 1' could be identified empirically (c.f. Slater 2001).

A related point also arises in connection with the application of *Reductio* in the above argument. For, under the influence of Priest in particular, it may seem that the application of this rule in connection with 'contradictories' is problematic. As Priest defined them in the initial quote from him above, 'contradictories' are merely 'things of the form α and $\neg\alpha$' (he meant, of course, things of the form 'α' and '$\neg\alpha$') and such 'contradictories' are therefore entirely syntactic objects, with the '$\neg$' in them possibly having quite different interpretations than Boolean negation. But the laws of logic apply not to physical signs themselves, but to their interpretations, and in particular *Reductio necessarily* applies if the symbol '$\neg$' in such forms as above is Boolean negation (c.f. Slater 2008a). If not the empirical letters, but their intended sense is attended to, therefore, the Millian conclusion above must hold.

4

But what about the Fixed Point Theorem? This is a theorem that holds in formal languages, where the sentences can all be numbered systematically in a list. Gödel first formulated such a numbering, and so any such is now commonly called a 'Gödel numbering'. Then what is provable in such languages is that, for any one-place predicate 'is P' in the language there is a sentence 'p' such that p if and only if the Gödel number of the sentence 'p' is P. So taking 'is P' to be the predicate 'is not the Gödel number of a true sentence', for instance, a mathematically demonstrable case of paradoxical self-reference is obtained assuming truth is taken to be a property of sentences. The question is: why cannot a comparable theorem be proved in the case of statements?

Here, in addition to the points made by Feferman and Koslow (which essentially follow the above point about the identity operator in the modal system KT), it is useful to remember Gödel's First Incompleteness Theorem, for instance, which showed that the undecidable formula '$\neg \exists x$ Prov (xg)' must have non-standard interpretations. Standardly this formula reads 'there is no number which is the Gödel number of a derivation of the formula with Gödel number g'. But Gödel showed this formula to be not derivable in the formal system under consideration (that of *Principia Mathematica*), if it is consistent, and that means that this same formula may be used to express truths, and express falsehoods, in connection with different, standard, and non-standard models. Of course, at the more mundane level of natural languages, there are many sentences that can be used to make an unlimited number of statements those with normal indexicals in them, such as 'now', 'then', 'here', 'there', 'this', 'that'. But such indexicality has been deliberately excluded in formal languages, and that has meant that the distinction between sentences and the statements they may be used to make has been largely lost, in the associated logical tradition. The result that Gödel proved, however, shows that a kind of indexicality is inescapable in languages that are sufficiently rich to accommodate a Gödel numbering of the sentences within them. And that shows that the categorical distinction between sentences and statements has to be retained. More specifically, it means that, while the sentences in some language might be numbered, the statements those sentences might be used to make are numberless. So that is fundamentally why there is no Fixed Point Theorem for statements: there can be no Gödel numbering of them.

Of course, if there are no paradoxical cases of self- (or cross-) reference with the relevant truth bearers, then many of Priest's arguments (also the arguments of many other people in the area) are proved invalid. But the way that Gödel's First Incompleteness Theorem forces a categorical separation between sentences and statements is particularly important in connection with an early, and very influential argument Priest gave for his Dialetheism in 'The Logic of Paradox'. For there he had said (Priest 1979, 271, see also 222-3):

> Thus we see that we establish claims of mathematics, if they are not axioms, by proving them (in the naive sense) from those axioms. This all seems obvious to the point of banality. However, it runs us straight into a major problem. For there seems to be no doubt that this procedure could be formalized. The axioms could, in principle, be written in a formal language and the proofs set out as formal proofs. The formal system that resulted would encode our naive proof methods. Moreover, there seems no reason to doubt that all recursive functions would be representable in the

system. For certainly all recursive functions are naively definable. However, according to Gödel's incompleteness theorem in any such formal system there will be sentences that are neither provable nor refutable at least if this set of axioms is decidable. If this were all there is to Gödel's theorem, the result might be surprising but not particularly worrying. The incompleteness of the formal system would merely show that there were mathematical problems beyond the powers of our proof procedures to settle. But this is not all there is. For some of these unprovable sentences can be *shown* to be true, i.e., proved in the naive sense. But the formal system was constructed in such a way that it encoded our informal proof methods. So there can be no such proof. Gödel's theorem presents an epistemological problem that has never been squarely faced. How is it to be resolved?

The error in this story occurs when Priest says 'some of these unprovable sentences can be *shown* to be true, i.e. proved in the naive sense'. For what is true is not the sentence '$\neg\exists x$ Prov (xg)' but the statement that no natural number is the Gödel number of a derivation of the formula with Gödel number g in the formal system in question. That statement is not the sentence itself, but instead is *what the sentence says on the standard interpretation*: $< \neg\exists x$ Prov (xg)$>$.

Forgetting this question of which interpretation is taken makes other people (e.g. for one, Williamson 1996) believe that humans might themselves be Turing Machines without knowing which they are, or, alternatively (see Penrose 1990), that humans might have some mysterious extra physiological facility in their brain, taking their production of true sentences beyond the capacity of any Turing Machine. Neither is the case: humans simply have the ability to give interpretations to sentences, and so to make, and prove statements. Priest's 'provability in the naive sense' is the provability of statements, while what he, with many others, would call 'the proof of a formula' is, instead, the derivability of that formula from certain others using some purely syntactic rules.

5

The basic problem with Priest's thought is therefore that he is a Formalist, and has not investigated the forms of speech that enable one to talk about *the content* of sentences. This reflects a much more general misapprehension which has been widely prevalent during the same period, and indeed for many decades before. In a large part, because the dominant approach has been Nominalistic during these times, it stems from not having an appropriate symbol for thoughts in professional discussions. Frege's 'horizontal', which was a kind of content stroke, was not copied in later developments of his logic, and only more recently has any other comparable symbol been promoted.

The abandonment of thoughts goes back at least to Hilbert's days, and, of course, it was not helped by the attack on propositions by Quine. But Hilbert's meta-mathematical programme was more fundamental, since that held, first of all, that it was all and only axiomatic structures that were the proper subject of foundational study. It is this study that has had the widest impact, not only in Mathematics but also in Logic. As Hilbert developed it, though, it had a singular difference in character from earlier studies of axiomatic structures. For Hilbert's approach was explicitly *meta-linguistic*, i.e. concerned just with the language, and formulae that appeared in the axioms. Objects satisfying the axioms have come, as a result, to be said to form a model of them, and reference to any such objects is in what was later called the 'object language', which must be contrasted with the 'meta-language' in which the meta-mathematical analysis is expressed.

Hilbert established the plausibility of his line of research with his axiomatisation of Geometry, which dispensed with Euclidean figures, and proceeded entirely by means of logic from completely explicit geometrical postulates. The removal of diagrams took foundational studies away from 'intuition' in the philosophical sense. More plainly, it takes one away from what the language in the axioms is about. As a result, despite wanting to say he had provided a foundation for 'Geometry' Hilbert had nothing to say about the lines and points in Euclid. Certainly the words 'line' and 'point' appear in Hilbert's axioms, but they were taken to apply simply to anything that satisfied the axioms. So the fact that those axioms did apply to Euclid's elements was quite incidental to Hilbert's interests, and remained something Hilbert did not attempt to provide any foundation for (c.f. McGuinness 1980, 40-1).

The basic error in Hilbert's programme was therefore that it gave no account at all of what is true in a model of some formulae, being deliberately concerned entirely with the formulae themselves. Hilbert considered the consistency of his formal systems to be very important, but it would require more than consistency to establish results about numbers from proofs in the kind of arithmetical formal system he was concerned with. There would need to be some proof of the *soundness* of that formal system, on the standard interpretation, before one could show even that 2+3=5, for instance. This follows from the character of Hilbert's meta-mathematics just in itself, it is important to note: there was no need to wait on Gödel, for instance, to point it out. Gödel's Theorems do not show, that is to say, that while some standard arithmetical truths are provable meta-mathematically, others are not. In fact none are, since any derivation within a formal arithmetical system must be supplemented with a demonstration of its soundness, on the standard interpretation, before any standard arithmetical facts can be proved on its basis. It is important to underline this, since current received wisdom seems to be that the faults in Hilbert's Formalism were only revealed once Gödel's Theorems were proved, whereas the present kind of error was there right at the start. The dominance of Gödel's Theorems in current thought about the Foundations of Mathematics presumably arose as a result of believing that, before Gödel proved his theorems, Hilbert's Formalism had no serious arguments against it — even though it is well known that Frege wanted Hilbert's formulae to express thoughts (c.f. McGuinness 1980, 48).

The grammatical point made before, about the difference between sentences on the one hand, and statements and propositions on the other, is crucial to seeing the detail of the needed correction to Hilbert. Sentences are mentioned using quotes, but when used (on the standard interpretation) they express propositions, and make statements, which are designated by the associated 'that'-clauses (McGuinness 1980, 164). The turnstile symbol in systems of formal logic and arithmetic is therefore mistakenly read, if it is read 'it is deducible that'. For the turnstile symbol is a meta-level predicate of sentences, whereas the reading then given involves an object-level operator on sentences. The contrast can be made even more sharp once one remembers the fact, like that about the truth operator before, that 'it is deducible that p' is equivalent to 'that p is deducible'. For the latter is of a subject-predicate form, and so the predicate 'is deducible' in it is an object-level predicate expressing a property of statements. To get from the meta-level predicate of sentences to the object-level property of statements one needs a proof of the soundness of the formal system involved, on the standard interpretation, and the processes involved in this additional matter of soundness have to be of quite a different character from any involved in the system in question. Indeed they cannot be formalistic at all, and might easily be called 'naive' — though therefore not entirely in Priest's sense.

Unlike a statement about a sentence, a statement about a statement is not about a purely syntactic form, i.e. some symbols independently of their meaning. But statements about sentences have dominated the Philosophy of Mathematics since Hilbert's meta-mathematics got its grip. So, clearly they have done so illicitly, because of the above points. Certainly meta-linguistic remarks about sentences have, quite properly, entered into the theory of computing, since a computer, of course, cannot take account of the meaning of any of the symbols fed into it. But the bulk of mathematics is not a meta-linguistic study of symbols, and is instead concerned with statements about other things: proving <2+3=5>, for example, rather than deriving '2+3=5'. Moreover, it is concerned with proving <2+3=5> *absolutely*, whereas derivations in a formal system are always relative to the axioms and rules that define that system. One might derive '2+3=5' from axioms 'A1',..., 'An', using rules R1,...Rn, but when proving <2+3=5> there is no such relativity. Of course a proof is involved in the formal case, since <'2+3=5' is derivable from axioms 'A1',..., 'An', using rules R1,...Rn> is proved. But what is then proved is not <2+3=5>.

The point shows that it is probably not an accident that most working mathematicians to this day (like Wittgenstein), give so little time to Gödel's Theorems. For, specifically, they are not relevant to the Foundations of Mathematics, if that is concerned, amongst other things, with the basis for what is true in the standard model of axiomatic arithmetics. For there are no fixed points with statements, as we have seen, and so no analogues of Gödel's sentential results can be constructed in connection with them. Hence these results are not relevant to *Arithmetic*, as it was conceived before axiomatic studies of uninterpreted formal systems came into vogue, and, with them, non-standard models of such structures. In addition, a full proof of the fact that 2+3=5, for instance, is not available from within them. Instead it can be drawn from such illustrations as the stick figure with five lines that Wittgenstein considers (Wittgenstein 1978, 58f). Only in a practical case like that, where the numerical terms are applied (and so are used and not just mentioned), does one get beyond numerals, and other symbols, and begin to work with their meanings. No string of sentences in a formal proof can get to anything in the right category, even; no computer can do so either, therefore, since no computer can give an interpretation to the symbols it processes.

The common convention of not showing quotation marks around formulae (as in the first quote from Priest at the start of this chapter) does not help people remember what it is that is 'provable' one should really say 'derivable', as before in a formal system. Only formulae are derivable, and, clearly, there is no 'proof', involving just a series of formulae, that Peano's postulates are true on the standard interpretation, for instance. For the expression 'that Peano's Postulates are true on the standard interpretation' is a noun phrase, and not a sentence, and so, *a fortiori*, it is not the last sentence of any rule-governed series of sentences. Neither, of course, can any arithmetical fact be in this position, since the noun phrase 'that 2+3=5', for instance, is equally not a sentence. The proof of the arithmetical fact this noun phrase refers to, therefore, has to be non-formal, at least at some stage, and can even proceed entirely in this way, as Wittgenstein has illustrated in several cases such as that above.

The use of physical objects is one thing that is distinctive about Wittgenstein's proof that 2+3=5 using a picture of five sticks grouped into a pair and a trio at one end, while all are collected together at the other end. Charles Parsons, likewise (Parsons 1979-80, 145-160, and 1986, Ch 11), has discussed such stick diagrams in connection with the use of 'mathematical intuition', and he has generated substantial portions of standard Arithmetic on this basis. But Wittgenstein's discussion does not go into such further details, and is consequently more basic and fundamental: it is concerned merely with the *foundations* of mathematics, in the proper sense of 'foundations'. Wittgenstein's discussion principally concerns the use of sticks, and the like, as *paradigms* — paradigms of countable things, for a start, and then of two things, of three things, of five things, etc. in the particular case above. Such paradigms help fix the normative criteria associated with the types of thing represented by the physical tokens in question. Books on Wittgenstein's Philosophy of Mathematics have not dwelt on these matters overmuch. But Frascolla is one commentator who addresses the required issues closely (Frascolla 1994, 145-60, see also Baker and Hacker 1980, Ch V). Frascolla discusses Wittgenstein's diagrammatic proof that the fingers of a hand and the vertices of a pentacle are the same in number. The incorporation of such physical paradigms into the language, in other cases, such as The Standard Metre, and colour samples, is a well-known part of Wittgenstein's later analysis of 'simples' (Fogelin 1976, 108f).

It is ironic in this connection that Gödel believed in 'intuition', even though he was so much a Platonist that he believed there was another world of abstract objects accessible to a specifically mathematical intuition. For the traditional philosophical description for the particular use of ostension in diagrammatic proofs, was that it was a matter of applying one's 'intuition' although that far more in the Kantian sense, which involved intuitions just of the spatio-temporal world, leading to synthetic *a priori* truths rather than trivially verbal, analytically *a priori* ones. What is also highly ironic is that the philosophical problems

Hilbert overlooked in his meta-mathematical programme have a formal resolution in the improved predicate logic he himself introduced the Epsilon Calculus
through its representation of Wittgensteinian 'simples' (Carnap 1961, see also
Psillos 2000). The place of such simples in mathematics has been said to be
Wittgenstein's later view of the synthetic *a priori*, the possibility of which
had been ruled out in the Tractatus (see Appendix 1). But it is possible that
Wittgenstein took a while to leave behind the more formalist account of mathematics he promoted in the Tractatus. Some of his own later remarks on Gödel's
Theorem, for instance, which Priest finds sympathy with (Priest 2004), betray
a continuing belief that any truths in mathematics are entirely analytic.

But it is another distinctive feature of the stick case that raises the most
serious questions about Hilbert's meta-mathematics. For Wittgenstein, in his
proof that 2+3=5, was clearly concerned with *the standard model* for the kind
of formal Arithmetics that were developed in the period. And, under the
influence of Hilbert's meta-mathematics, proofs in the standard model were
eschewed, in favour of meta-linguistic derivations of formulae, independently of
their interpretation. Formal Arithmetics dealing with such derivations crucially
have non-standard models as well as standard ones, as we have seen. But the
problem with any model, from the meta-mathematical perspective is that it
cannot be specified formally. Again, some kind of 'intuition' is involved in
grasping what such a model is (c.f. Dummett 1978, Ch 13).

This is the sticking point for Formalists: acknowledgement of the kind of
foundational issues that Wittgenstein, notably, as a primary teacher at Trat-
tenbach and neighbouring villages, would have been only too conscious of. He
would have been only too conscious of teaching 'the sorts of things that children become familiar with when they learn to count and to do arithmetic'
(Priest 1979, 221). Of course, it is not just children that are in need of informal elucidations of the required sort; beginners of all ages are in the same
position, as teachers of elementary logic courses at tertiary level, for instance,
will be only too well aware. Getting students to realize to be true things of the
form <'2+3=5' is derivable from axioms 'A1',..., 'An', using rules R1,...Rn>
is not something which can itself be formalized in terms of the derivation of
one sentence from another. Competently producing and comprehending such
derivations is a learned art, but neither the teaching nor the learning of that
art is itself an exercise of it. Forgetting such 'naive' matters, and especially
the then learned difference between '<p>' and ' "p" ', is what has centrally
led to all sorts of trouble with sophisticated twentieth century logic and the
philosophy of mathematics, as we have now seen.

CHAPTER 8

NATURAL LANGUAGE CONSISTENCY

1

Tarski's assessment that natural language is inconsistent on account of the Liar Paradox is shown to be incorrect: what Tarski's theorem in fact shows is that Truth is not a property of sentences but of propositions. By using propositions rather than sentences as the bearers of Truth, semantic closure within the same language is easily obtained. Tarski's contrary assessment was partly based on confusions about propositions and their grammatical expression. But more centrally it arose through blindness to pragmatic factors in language — a blindness that was common in his time, and which has continued to the present day, in discussions of 'Open Pairs', and Yablo-type paradoxes, for instance. For completeness, it is also shown that the Fixed Point Theorem does not apply to propositions, because of categorical differences between sentences and propositions.

2

Natural language has resources that have not been copied into the formal languages of recent logic, and in at least two cases this has led to near intractable difficulties in that discipline. There are no reflexive pronouns in the languages of recent logic, which has produced misrepresentations of notorious predicates like 'is not applicable to itself' and 'is not a member of itself' (Slater 2004b, 2005b, and Appendix 2); and there are no nominalising devices in the languages of recent logic, which has engendered the abandonment not only of 'that'-clauses, but also their referents: propositions. It is the latter difficulty that is the main focus of the present chapter, although together the two difficulties have combined to generate what have been called 'the paradoxes of self-reference', one consequence of which has been the judgment that natural language is inconsistent, on account of such things as the Liar Paradox.

This view of natural language, of course, is a mirage brought about by the above removal of standard features of ordinary speech in the restricted languages that are instead considered within the formal tradition. But it needs to be shown in some detail how it can be that natural language, by retaining these features, does resolve the classic self-referential paradoxes. The first point to realize is why there is no direct analogue of the Fixed Point Theorem in connection with propositions. For it is that theorem which seemingly guarantees, in a very rigorous formal way, that there is no 'semantic closure', i.e. no consistent truth predicate both applicable to and definable within the same language. But only a glance at it is sufficient to show that it has no direct analogue with propositions. For it has the form:

$q \equiv A(gn'q')$,

and equates certain propositions with propositions about the Gödel numbers
of sentences that express them. So the only point that needs to be made with
respect to natural language is the simple one that the propositions expressible
in a language, unlike the sentences in that language, cannot be numbered.

The point is obvious, given that natural language contains explicit indexicals.
But we shall see that it holds much more widely. Thus while there is one
sentence 'this is not true', innumerable propositions may be made with it,
depending on the chosen referent of 'this'. And that multiplicity of propositions
not only defeats any direct analogue of the Fixed Point Theorem applying, it
also immediately resolves the supposed self-referential paradox that might be
formulated in this case. For if the referent of 'this' is taken to be the sentence
'this is not true' itself, then the proposition then made by the sentence is simply
true. It is true that the sentence 'this sentence is not true' is not true, because
truth attaches to the proposition it makes in the given circumstances, instead.
Only a failure to separate the sentence itself from the proposition it is chosen
to make could result in the belief that the sentence both was not true, and was
true, so the contradiction is immediately avoided once the proposition and the
sentence are separated. Of course that requires distinguishing a mentioned, i.e.
quoted sentence from a used sentence preceded by the nominaliser 'that', and
more will be said about that in due course.

Before that it is important to see that what might still be called the 'indexicality' of language extends much further than in such explicit cases where a
demonstrative like 'this' is involved. Perhaps the term 'contextuality' would
better convey the general idea in the wider context, but it is still the same
feature that is present in explicit indexicals.

To see that wider presence of the required features we must see, pre-eminently,
the contextuality of what Quine called an 'eternal sentence' like 'the sentence
at the top of page n of book B is not true'. For it would appear that the contextuality of such 'eternal sentences' is what has primarily been missed, since
the location of (some token of) such a sentence (at the top of page n of book
B) might be taken to be a plain fact about the actual world, and so seemingly
not one in a special, limited context. Hence, it would seem, the sentence alone
could be taken to carry the appropriate truth-value, though of course, then, a
very puzzling one. But in this case the variability of truth-value of the associated propositions is with respect to *different possible worlds*. Here is where
an historical point gains some significance: it is not an historical accident that
the self-referential paradoxes arose forcefully in logicians minds before context
dependence and indexicality were studied fully, for what was also characteris-
tic of that period was that *possible worlds* were not studied either. Indeed, if
one keeps to that early twentieth century mind-set one will still find the self-
referential paradoxes puzzling, particularly followers of Quine on Modality, and
eternal sentences, of course, since one needs to understand fictional contexts
in order to realize that the actual world is just another context. Contrariwise,
moving out of that mind-set, and in particular moving away from the influence
of Tarski, the resolutions of such paradoxes become extraordinarily easy, as
above. For what is true in the given 'eternal sentence' case is simply that the

sentence at the top of page n of book B is not true, while the key point to realize, to avoid contradiction, is that to say that, i.e. to say that that proposition
is true is not to say that the sentence itself is true. What is true is not the
sentence, but what the sentence says on a self-referential interpretation.

The crucial point is that it is not possible for a referential sentence to specify
which world it is used in, since that is a matter of pragmatics. Even if it was
possible that a code might be provided giving the various referents, in different
possible worlds, of phrases like 'the sentence at the top of page n in book
B', etc., still what world the phrase is used in is a matter of pragmatics. So
the code itself shows that such phrases are trans-world, if not 'contextually'
indexical, and so ambiguous, along with standard referential indexicals like
demonstratives, and pronouns. In the general case of the basic type of puzzling
syntactic identity

  t = 't is not true',

the grammatical point is that 't' is quoted on the right hand side, and so is just
mentioned without reference to a world. So no specific referent can be involved
there. But the referent of 't' *in this world* is involved on the left hand side,
because it is not mentioned but used, and we are speaking in this world. But
if one tries to improve on this by trying to insert the context of utterance into
the sentence itself, constructing something like

  t = 't when uttered in this world/in the actual world is not true'

then one either has brought in an explicit indexical ('this world'), or an equivalent to one — remembering David Lewis on 'the actual world' (Lewis 1986).

Of course, to see how one can use referential phrases 'in other worlds' one has
to remember also that other possible worlds are not entirely abstract objects,
since we can imagine entering them, which is a process that takes place in this
world. So there is no difficulty in transferring oneself, in one's imagination if
not in reality, to another possible world, or situation. In the case of linguistic
fictions this commonly involves certain context markers, like 'Once upon a
time', for instance. Maybe once upon a time an old man was reading page n
of a book, B, the first sentence on that page [i.e. page n of book B] being 'The
first sentence on page n of book B is not true'. In this case, at its unquoted
place the subsequently quoted referential phrase refers to the page the old man
was reading in the fiction, i.e. the possible world. Without the 'Once upon a
time', the story starting 'a man was reading' might be fiction or non-fiction,
although the same kind of linguistic cross-reference would still occur, from the
referential 'the first sentence...' back to the previous introductory description
of the old man. In non-anaphoric uses of referential phrases, i.e. when they are
'deictic', there is no context marker like 'Once upon a time', or even explicit
introductory description. But, now, *the absence of such a context marker* is
not part of the sentence(s) that follow, so, even when there is direct reference
to the actual world, that is a matter of the pragmatics, not the semantics of
the utterance, and therefore not something in the sentence alone, in itself.

Wouldn't there still be a paradox if we considered whether the sentence token

  This sentence token is not true

was true or false? Here we, in this world, are referring to a specific set of words
also in this world, and there is therefore no possibility of ambiguity of reference.

But, still, what is true is *that that sentence token is not true*, not the sentence token itself. It is the proposition expressible by the sentence token which is true, while the sentence token itself is not true.

3

The above points about the categorical differences between sentences and propositions show up in a variety of individual ways in connection with further particular discrete cases in the area. Indeed, the paradoxes where indexicality is showable by direct inspection of the sentences, along with elementary cases involving less plain and evident indexicals, like 'the first sentence on page n of book B is not true', clearly give good inductive grounds for the belief that there is indexicality in all related cases people have found paradoxical. Note that supposed examples of necessarily self-referential sentences, such as 'this very sentence is false' do not escape from the indexical category, since in fact the referent of the 'this very sentence' still has to be determined with a gesture, and might be to some other sentence, so the 'self-reference' is not properly necessary. If one gives names to sentences there is the same problem as we shall see arises with numbering systems, since it is not in 'sentence A is not true' itself that it is sentence A in some list, if it is so, and that very same sentence might make a different proposition using a different naming system. It is the resulting difference between sentences and propositions that resolves a number of further paradoxes.

Recently, for instance, there has been much discussion about what are now called 'open pairs' and the like (e.g. Sorensen 2003, Armour-Garb and Woodbridge 2006). Here are some sentence-proposition facts regarding what are often presented as

1: 2 is not true,
2: 1 is true.

They can be extended quite easily to other cases, such as 'open pairs' themselves, and also Yablo's (Yablo 1993, see also, for instance, Goldstein 2006). Maybe the idea is that the examples given can be expunged of contextual elements, and turned into 'eternal sentences' (the matter dealt with before), but as they stand these cases even more clearly do not give rise to any paradox, once the very evident contextuality is spelt out. For, if '1' and '2' are supposed to be names of sentences, then, for a start the proper, full expression is quotational:

Sentence 1='Sentence 2 is not true',
Sentence 2='Sentence 1 is true'.

The puzzle then seems to be that if sentence 1 is true then (because of what sentence 1 says) sentence 2 is not true, but that requires, doesn't it (because of what sentence 2 says), that sentence 1 is not true? Hence, seemingly, sentence 1 has to be not true. But that makes sentence 2 true, surely (because of what sentence 1 says), and therefore we seem to be able to deduce that sentence 1 is true (because of what sentence 2 says)? But once we remember the sentence-proposition distinction it becomes clear that the sentences on the right are neither true nor false in themselves, indeed, in themselves they are not the sort of thing that can have truth-values. To gain a truth-value they need to be fully interpreted, which in this case means they need to be used in connection with a list of sentences in a numerical order, as on the left. Then what proposition sentence 1 can be used to make in the given context, namely that sentence 2 (here) is not true, is true, showing one must distinguish very carefully, again, the sentence from what it says in the context of the given list, i.e. what proposition it makes there. For sentence 1 is not true, while the proposition made by it, in this context, is true. And likewise, in reverse, with sentence 2, since the proposition that it makes is that sentence 1 (here) is true, which is not true. So not only sentence 2, but also the proposition it makes in the context is not true.

As before, this kind of point must also be made in the central, 'self-referential' case, where, for instance,

Sentence 3='Sentence 3 is not true'.
For here what is true is not (contradictorily) sentence 3 but the proposition made by it in this context, namely that sentence 3 (here) is not true. Hence there is no Liar paradox in this case, and it is the functioning of the pragmatic context that has been primarily overlooked by theorists who find further paradoxes in this area, i.e. those who find Open Pairs and Yablo-type cases puzzling.

Notice, in connection with 'that'-clauses, that one cannot just nominate a self-referential propositional identity, in the same way as a 'self-referential' sentential one. There is no barrier to naming the sentence 'sentence 3 is not true' as 'sentence 3', since the identity of the sentence is independent of what the phrase 'sentence 3', in it, refers to. Not so with any attempt to name, for instance, the proposition that proposition 3 is not true, as 'proposition 3' (c.f. Kripke 1975 note 5). Thus what people have in mind with

3: 3 is not true
might be not a sentential identity but a propositional one:

Proposition 3 is that proposition 3 is not true.
But to know the identity of the proposition on the right hand side of this identity one needs to already know the referent of the 'proposition 3' used in making that proposition. So its referent is settled, and one is not free to nominate the referent of the expression subsequently. Of course, no referent of the expression is given by the identity as stated, since that is circular and so leads to an infinite regress, as Kneale pointed out (Kneale 1972, 242). Yet another way of realising this point is by seeing that substitution into the right hand side of the previous sentential identity is impossible, because of the quotation, whereas substitution into the right hand side of the propositional identity is possible, because of the lack of quotation. The same point holds with other phrases referring to propositions, e.g.

What sentence 3 says is that what sentence 3 says is not true,
although care must be taken to distinguish this from the unproblematic:

What sentence 3 says is that sentence 3 is not true.
The general, overriding point is that there is no syntactic self-reference, since any reference is only given through an interpretation. Thus one must primarily remember, for instance, that a sentential identity like

t= 't is not true',

does not itself show that some sentence is about itself, since it does not entail that

$(\exists x)(x = $ 'x is not true'),

by Existential Generalisation. By contrast, if one talks not about the identity of a sentence but about the content of it, in a sentence such as

t says that t is not true,

there follows unproblematically that

$(\exists x)(x$ says that x is not true),

and so that something is self-referential.

The basic confusion, therefore, is a use-mention confusion and it is primarily separating clearly use from mention that shows there is no problem with any version of the Liar Paradox — or Open Pair, or Yablo type case, as above. For it is only on a certain interpretation that paradoxes arise, and it is now clear that that is an additional, intensional matter beyond any extensional, direct speech identity. The fact that the interpretation is an intensional matter is what brings in the need for indirect speech, and 'that'-clauses. But that then shows that what is relevantly true in paradoxical cases is simply that some sentence is not true, while also showing that to say that some sentence is not true is not to say that what is true is some sentence. What is true is what some sentence says on a given interpretation, i.e. the proposition it makes in those circumstances.

4

But the pragmatic dimension involved in 'that'-clause use extends well beyond indexical sentences, eternal sentences, and listed sentences. It reaches even the area of mathematical sentences, which at one time were thought to be immune to truth-value variability. For, while it was maybe quite plausible to believe that one can have a truth predicate of elementary mathematical sentences such that, e.g.

T '2 + 3 = 5' if and only if 2 + 3 = 5,

the realisation that the cases where this kind of equivalence has a chance of holding are very special cases has now begun to dawn — given Tarski's own theorem, showing that there cannot be such a 'T' in general, even in Arithmetic. If there are alternative models for arithmetic sentences, as Gödel demonstrated, then no formal equivalence like the one displayed is available, in general, and only *the use* of the sentence on the left in connection with the standard model for Arithmetic would produce a necessary *propositional equivalence*,

It is true that 2+3=5 if and only if 2+3=5.

But that is an instance of Horwich's Equivalence Scheme (Horwich 1998), not of Tarski's Truth Scheme (Tarski 1956).

Quite a lot hangs on this, as we have seen. What is true, in Gödel's First Incompleteness Theorem, for instance, is not the sentence '(x)Fx', for a certain predicate 'F', but the proposition that all natural numbers are F, i.e. the proposition expressed by the sentence when it is used with respect to the standard model. The inability of the logical tradition to represent such a propositional referring phrase as 'that all natural numbers are F' has, by contrast, made it seem that what is true or false on the standard interpretation is still the (mentioned) sentence '(x)Fx'; but only that sentence in a certain use, preceded by 'that', refers to the item that has the truth-value. For the formula '(x)Fx' is indexical, because the universe of discourse of the quantification is variable, and that allows different propositions to be made with this same sentence, while only one such proposition is claimed to be true.

The major consequence of this, not always drawn, is a very large one indeed: that humans are categorically different from Turing machines. For while, like Turing Machines, humans can utter sentences such as '(x)Fx', they can also do something Turing Machines cannot, namely use sentences like '(x)Fx' to state things about different models. In particular, a Turing Machine would have to not only utter '(x)Fx', but also use it pragmatically in connection with one model rather than another, and specifically in connection with the standard, intended model, if it was to state that all natural numbers are F. But a Turing Machine lacks the required power of choice to select the standard model, and thereby any capacity to prove that Fn for any natural number n, even if it can generate the sentence 'Fn' for every numeral 'n'.

The central point is that Tarski, although he continuously expressed propositions, and made statements, was not conscious in a theoretical way either of the existence of propositions and statements, or their distinctive grammar. But, of course, the idea that sentences are the bearers of Truth was not just Tarski's opinion. There were many reasons advanced by theorists, in the early decades of the twentieth century, for the abandonment of statements and propositions, and concentration instead on sentences as the bearers of semantic assessments. More important in the present context was the main practical measure to the same effect: the preference for an unnatural language in which such abstract objects could not even be referred to, or talked about. The principal kinds of expression that do that, in natural language, are the 'that'-clauses focussed on above. 'That'-clauses are substantival phrases such as occur in subject-predicate sentences like 'That the Kneales showed how to refer to properties and propositions is true, but not well known' (see, for instance, the O.E.D. under 'that' as a conjunction). Intensional Logic, as currently developed, deals with related expressions. It deals with the 'cleft' form of such sentences, i.e., in the case illustrated, 'It is true, but not well known, that the Kneales showed how to refer to properties and propositions'. But that would be symbolised as involving the operators 'it is true that', and 'it is not well known that', and so it would be expressed in a language in which 'that'-clauses have no distinct, substantival place.

Difficulties with the recognition of propositional referring phrases of the form 'that p' are therefore a large part of what have made Liar sentences seem paradoxical. Frege's content stroke, i.e. the horizontal line that he sometimes used to indicate the thought expressed by a sentence, has not been incorporated into the generality of logic texts that have followed his formal work, and that has caused many of the difficulties (Slater 2001, and chapter 4). Using a quotational form 'p' as an alternative to 'that p', as is commonly done, confuses syntactic expressions with their semantic and pragmatic readings, and leads to misunderstandings about the differences between Tarski's Truth Scheme and Horwich's Equivalence Scheme.

One difference between Tarski's Truth Scheme and the Equivalence Scheme of Horwich, for example, is that only the latter applies to indexical cases. Horwich's is a propositional schema, viz

the proposition that p is true if and only if p,

whereas Tarski's is a sentential schema:

the sentence x is true if and only if p,

where what replaces 'x' is a name of a sentence whose translation into the metalanguage replaces 'p'. The difference is most pointed in the homophonic sentential case, which parallels very closely the propositional one. For what replaces 'x' then is a quotation-name of the sentence that replaces 'p', not that sentence itself. So one could have, for instance,

that he is happy is true if and only if he is happy,

while one cannot have

'he is happy' is true if and only if he is happy.

Certainly there would be less need to make the distinction if all sentences were unambiguous and non-indexical, i.e. had just one interpretation, since then facts about propositions could be mapped 1-1 onto facts about sentences. But the central question, as we have seen, is whether sentences are non-indexical in the required way.

As was pointed out before, the difference between 'that p is true', and " 'p' is true" may not be completely appreciated even in Horwich's informal work, since he thinks there are still paradoxical cases of his propositional schema. But 'it is true that' is the null modality in the modal system KT, i.e. an 'L' for which it is necessary that $Lp \equiv p$, and so one cannot have that $p \equiv \neg Lp$, since the modal system KT is consistent. Horwich's thought against this, as we saw, is in terms of 'THE PROPOSITION FORMULATED IN CAPITAL LETTERS IS NOT TRUE', and we can now look in more detail at the argument he had in connection with it. The sentence in question he abbreviates to '# ' (Horwich 1998, 40-41). He also, of course, quite generally abbreviates 'the proposition that p' to '<p>'. So how does he deduce that '#' is paradoxical? Naturally he gets '# ' from '<# > is true', and '¬# ' from '<#> is not true'. But he also wants to derive '<#> is not true' from '#' ('whose subject, said to be not true, turns out to be the proposition <#>') and '<#> is true' from '¬#' ('which says of <# > that it is *not* not true'). So he ends up saying that <#> is true if and only if <#> is not true.

There clearly must be something wrong with this argument if 'It is true that p' is equivalent to 'That p is true' and the former is quite consistent. But where does Horwich go wrong? He goes wrong through thinking that a context-independent proposition is expressed in the case in question. Specifically, what the contradiction shows is that the referring phrase 'THE PROPOSITION FORMULATED IN CAPITAL LETTERS' must be indexical, allowing '#' not to express a definite proposition on its own any more than the 'This is not true' does, on its own. But an indirect proof of this fact is hardly needed, since there are, of course, many propositions that are, or could be expressed in capital letters.

A more commonly presented example, requiring much the same kind of solution, arises with, for example,

    The proposition expressed by this sentence token is not true,

when the referent of 'this sentence token' here is supposed to be the sentence token just indented (which has to be added, since a token of the very same sentence, of course, could make reference to a quite different sentence token). But there is no paradox here, since that would arise only if a specific, single proposition was expressed by the sentence token in question. It was Prior who, in recent times, first came to suspect, in connection with such cases, that there was the possibility of ambiguity, preventing a single proposition being expressed. Thus he said (Prior 1971, 106): 'We could then say that if x means that x is false it will have two contradictory meanings that it is false and that it is true'. But the exploration of such possibilities arose even earlier, in the work of Thomas Bradwardine. See for complete details Read 2010.

CHAPTER 9

A PERFECT LANGUAGE?

1

In recent Formal Logic, languages without indexicals have been widely studied, since they have been thought to be more 'perfect' than our natural language. This chapter shows the error in this line of thought as well as providing a plausible account of why it has been so attractive to its formalistic adherents — it provides a central control, making what is on the page before one seem to be all that needs to be attended to. The possibility of a language without indexicals has been important, as well, to recent theorists about Truth, such as Tarski; and its attraction is shown to be a large part of the motivation for later developments in this view of Truth made by Priest. By considering languages containing indexicals instead, it is shown that it is the removal of them that has created many of the major problems that have arisen within this recent semantic tradition. Furthermore, one consequence of Gödel's first Incompleteness Theorem is that indexicality is inescapable in languages of sufficient complexity.

2

Some people might try to replace the indexical 'He is happy', said of John, with 'John is happy', to get a sentence with a fixed truth value, which might fit into the Truth Scheme. But the tense in 'John is happy' makes this sentence indexical another way. So Quine only held that 'He is happy', said of John *at time t*, should be replaced with the *eternal sentence* 'John at time t is happy'. If Quine had accepted Modal Logic, therefore, 'He is happy', said of John at time t in possible world w, might have been replaced with the eternal and transworld sentence 'John at time t in possible world w is happy'. But Quine's case is questionable, since the number of words in a language must be denumerable, while the number of times could be non-denumerable. And the possible-worlds case shows even more clearly that the Quinean procedure cannot be completed, since there is not even an ordering, let alone a numbering of all possible worlds. So one cannot always find a sentence with a fixed truth value for the Truth Scheme to apply to, i.e. get 'central control' *within the language* of what is being said. Indeed, can one even say: 'John at time t is happy' is true if and only if John is happy at time t? Which person named 'John' is being spoken about? Clearly the tradition has been presuming that any name in a sentence refers unambiguously just to one thing. But language cannot work like that: there are just too many things around at the same time, so names themselves have to be indexical.

Tarski had well-known problems with the Liar Paradox, but he also had problems with indexicals, since he thought that his Truth Scheme still worked when they were in the sentences considered. Thus Carnap, in his 'Intellectual Biography' says: I wondered how it was possible to state the truth-condition for a simple sentence like 'this table is black'. Tarski replied, This is simple: the sentence 'this table is black' is true if and only if this table is black. But the Truth Scheme " 'p' is true if and only if p" does not apply when indexicals are involved, since one cannot say: 'He is happy' is true if and only if he is happy. Even Horwich, who has studied the matter more than most, tries to keep indexical declarative sentences (along with all others exhibiting ambiguity and context sensitivity) within the purview of 'the disquotation schema'. He does so by inventing a device, which with 'this table is black', for example, would make 'this table', when merely quoted on the left hand side of the schema, have the same referent as when unquoted on the right hand side. But it is not the *mentioned sentence* 'this table is black' that is then said to be true, it is the *proposition* it makes at the place where it is *used*. It is not that 'this table is black' is true if and only if this table is black, but that *that this table is black* is true if and only if this table is black. Horwich has in this way been led to think there are closer parallels between sentences and propositions than in fact there are, and in particular he has been led to believe that propositions fall foul of Liar-type paradoxes, just like sentences. But we shall see they clearly do not, once the relevant distinctions between sentences and propositions have been made entirely clear.

The 'perfect languages' of Modern Logic have no representation of 'that'-clauses, but as we have seen, in natural language we make the above distinction between mention and use clear by separating out the sentence when quoted from the associated 'that'-clause. We distinguish the sentence 'snow is white', from the proposition that snow is white, and so say not that 'snow is white' is true if and only if snow is white, but that that snow is white is true if and only if snow is white. And once we do this then each of Tarski's problems mentioned before vanish. For if one replaces the quoted 'p' in Tarski's Truth Scheme with the related 'that'-clause, then, first of all, one gets Horwich's Equivalence Scheme: That p is true if and only if p. But that means we get something that applies to all declarative sentences — indexical sentences as much as to others — since all words, not just 'snow', but even words like 'this table', 'he' and 'John' can be used, in comparable positions, on both sides of this equivalence. But also the Liar Paradox is now seen to arise only with sentences. For, yet again, Horwich's 'That p is true', while it is of the subject-predicate form, is equivalent to 'It is true that p', which is of an operator form. And 'It is true that p' involves the operator 'it is true that' which is the null or identity operator in the modal system KT, which is consistent.

3

The Logical Empiricists, from whom much of modern Formal Logic derives, were not conscious of the use of 'that'-clauses as referring expressions. Indeed the tradition following them was taught by Quine to be ontologically nervous of what such expressions might refer to. But what such clauses refer to are things in everybody's everyday experience, so only by returning to mind everyone's everyday talk can one dispel any feeling of foreignness. The first thought of any

Logical Empiricist reading the present sentence, for instance, is most likely that its central claim, that it cannot be formalised in standard logic, is nonsense. For one can certainly formalise

(1) It is nonsense that the sentence above cannot be formalised in standard logic.

in standard logic. That is a straightforward intensional locution of the form 'It is nonsense that p', which could be symbolized 'Np', with 'N' the operator 'it is nonsense that'. The trouble with this first-thought response, however, is that it does not treat all the locutions in the sentence above — or even any exact one of them. For there are at least four others of note, viz:

(2) Its central claim is nonsense.

(3) That the above sentence cannot be formalised in standard logic is nonsense.

(4) The first thought of any Logical Empiricist reading the above sentence is most likely that its central claim, that it cannot be formalised in standard logic, is nonsense.

(5) Its central claim is that it cannot be formalised in standard logic.

If one checks English grammar books then one will find that (3) is equivalent to the intensional locution (1). The locution (1), involving the operator 'it is nonsense that', is merely the 'cleft' form of (3), with a dangling pronoun 'it' at the start waiting on the 'that'-clause that follows. But the re-expression of (1) in the exact form of (3), cannot be standardly symbolised, since there is no standard formalisation of 'that'-clauses as subject terms, or even as separate expressions in their own right. Using Horwich's symbol '<>' one could write (3) in the form 'N' <p>', with 'N' <>' being equivalent to the operator 'N' introduced before, but with 'N'' being the predicate of 'that'-clauses 'is nonsense'. In these terms (2) has the form 'N'r', i.e. it is a simple subject-predicate expression, with 'r' as 'its central claim'. The trouble deepens for Logical Empiricists, however, once one considers (4) and (5). For there is no rephrasing of these as 'intensional locutions' in the normal sense. Thus (5) is of the relational form 'r = <p>', with 'r = <>' being 'its central claim is that'. And 'it is most likely that the first thought of any Logical Empiricist reading the above sentence is that' is 'P(t = <>)', with 'P' the probability operator 'it is most likely that', or, equivalently, 'P' <t = <>>', with 'P'' the predicate of 'that'-clauses, 'is most likely'. The problem for Quineans is particularly acute because of this, since both (4) and (5) are revealed to involve a proposition being identified.

As one can check in the grammar books, a 'that'-clause, as object, usually follows the verb, e.g. 'I have heard that he was there'. As subject it is also most commonly placed after the verb but introduced by a preceding 'it', e.g. 'it is certain that he was there' = 'that he was there is certain'. The formal study of operators, such as those in 'it is certain that p', 'it is necessary that p', however, has been continuously pursued without any recognition of their predicative equivalents, as in 'that p is certain', 'that p is necessary'. A famous fixed-point theorem of Montague's showed operators cannot have predicative equivalents whose subjects are sentences; and a string of further, more fine grained results have followed detailing just how close to a predicative equivalent one can get, always supposing the correlative subjects are sentence-like. But what these authors are primarily forgetting is that there are exact predicative equivalents supposing only that the correlative subjects are 'that'-clauses. So the (restricted) mathematical equivalences they have gone on to derive fall merely into the category of recreational mathematics. Equivalences with predicates of more appropriate subjects are quite unrestricted, and with them mathematics is not involved at all in their proof, just elementary grammar. From his result Montague concluded that 'virtually all of modal logic must be sacrificed' if necessity is conceived of as a predicate of sentences. But the logic with which he was working contained no representation of 'that'-clauses as syntactic units, and not everything in modal logic needs to be sacrificed if necessity is conceived of instead as a property of propositions (though somethings need to be sacrificed on this and other grounds as we shall see in chapter 11).

The most important new form of expression that comes into logical theory as a result of having referring terms to propositions are propositional identities of the form 'x = that p'. There are many such with names in place of 'x', as we have seen. Quite commonly, however, the present generation of logical theorists have tried to insert a variety of set-theoretic expressions in place of 'x'. But propositions are not sets of any kind, as the bad grammar of the resulting identity statements of the form 'the set of Ss is that p' readily shows. A different language is needed from the set-theoretic one, extending normal formal ones. Using that extension it is then easy to symbolise, and validate many intensional inferences in natural language that previously have been left unformalised. The case we considered in chapter four was that

My proposal is that we go to Benidorm for our holiday,
with
You accept my proposal,
entails
You accept that we go to Benidorm for our holiday.
But there are plenty of other cases. For instance:
My first thought was that there had been an intruder in the house.
with
That an intruder had been in the house proved true,
entails
My first thought proved true.
Here again the argument cannot be expressed using operator constructions. Certainly the second premise can be re-formulated as one, viz:
It proved true that an intruder had been in the house.
But the conclusion cannot be put into operator form, even though 'It proved true, my first thought' is, of course, perfectly grammatical.

Other problems in this area, both particular and quite general, have been prominent in twentieth century logic and philosophy. Donald Davidson, for instance, proposed a 'paratactic' analysis of indirect speech, in his paper 'On

Saying That'. Thus he argued that 'Galileo said that the earth moves' should be analysed as 'Galileo said that' (with 'that' a demonstrative) and 'the earth moves'. But 'It is certain that he was there', as before, is the same as 'that he was there is certain', and more generally 'that'-clauses may occur as grammatical subjects as well as objects. So propositional identity statements like 'What Galileo said was that the earth moves' were not acknowledged in Davidson's discussion of 'saying that'. This one adds to 'Galileo said that the earth moves' a uniqueness clause, enabling identification of the only thing that Galileo said (in a given context). And it then licenses hitherto unconsidered inferences such as that from 'No one believed what Galileo said' to 'No one believed that the earth moves' and *vice versa*. Indeed, closer to Davidson's discussion of 'same-saying', there is also the fact that Davidson's central case, 'Galileo said that the earth moves', with the identity statement 'What I said was that the earth moves' yields 'Galileo said what I said', which cannot be given a 'paratactic' analysis. So Davidson's division of 'Galileo said that the earth moves' into 'Galileo said that' and 'The earth moves', was well off the mark, making it very puzzling why it was thought even worthy of consideration for so long.

But why ever were 'intensional constructions', as commonly understood, considered for so long? The central logical tradition since Frege and Peirce in the late nineteenth century has had three broad parts, it will be remembered. There is Propositional Logic, dealing with inferences involving such sentence connectives as 'and', 'or', 'only if' etc. One result here, for example, is *Modus Ponens*: that from 'p', with 'p only if q' one can derive 'q'. Then there is Predicate Logic, which is a development of the Syllogistic that Aristotle first formulated. That deals with inferences deriving from the structure of sentences containing quantifiers like 'all', 'some' and 'no'. One elementary result here, for example, is the Syllogism in Barbara: that from 'All A is B' and 'All B is C' one can derive 'All A is C'. These two parts of twentieth century logic have been very fully worked out. But there is a third part of this tradition, called 'Intensional Logic', which has had a very different history. Many of it's difficulties derive from forgetting aspects of the functioning of the term 'that' as above.

If one looks at standard texts on twentieth century logic then this third area that logic is taken to deal with covers sentences like 'it is possible that p', and 'it is necessary that p', which arise in Modal Logic. Also involved are sentences like 'A knows that p', 'A believes that p', which are dealt with in General Intensional Logic. The problem the tradition has had with 'intensional constructions' of these sorts started principally with some very influential writings by Wittgenstein in his *Tractatus Logico-Philosophicus*. For Wittgenstein asserted that the general form of a proposition was a 'truth function' of elementary propositions, and while this was easily seen to be the case in connection with Propositional Logic, and very plausibly the case in connection with Predicate Logic, it was not seen how this could apply to any extent in connection with Intensional Logic. For 'It is possible that p' and the rest are not truth functions of 'p'. The first attempt at an analysis of these forms, notably by Carnap and Quine, offered sentential theories of intensional constructions. Thus 'A believes that p' was to be understood to be equivalent to something like 'A believes-true 'p'', where 'p' was some sentence that expressed the proposition that p. That would allow 'A believes that p' to be dealt with by Predicate Logic, with 'believes-true' being a relation between two subjects 'A' and " 'p' ". One argument historically used against this analysis (by Church) was that there cannot be a strict equivalence between the two forms 'A be- lieves that p', and 'A believes-true 'p'', since the relation is dependent on the contingent assumption that the sentence 'p' does express the proposition that p. So the tradition, in the late 1950s, came to believe that 'A believes that p', along with all other intensional constructions was to be understood as involving an 'operator' rather than a predicate. Thus, for instance, the sentence 'It is possible that p' came to be analysed as involving the operator 'it is possible that' with, not the *mentioned* sentence 'p', but the *used* sentence 'p', i.e. 'p' without quotes, as its complement.

But why not construe 'It is possible that p' as 'That p is possible' and so as involving a predicate of the subject 'that p'? Why not construe 'A believes that p' like 'That p is believed by A', and so as involving a relation between the two subjects 'A' and 'that p'? Each case would then be handled in the standard manner of Predicate Logic, and there would be no breach of truth- functionality. There was quite a debate about construing 'A believes that p' as a relation, with Quine, Davidson and Prior arguing in different ways, but what is very noticeable in retrospect is that none of the prominent theorists of this earlier period (except Kneale) were at all conscious of the full use of 'that'-clauses as referring expressions, and so no attention was given to the variant form 'That p is believed by A'. Likewise with 'That p is possible'. But there are too many constructions in which 'that'-clauses are present *and which cannot be analysed in terms of operators* for them to be ignored for ever.

4

As we have seen, attention to 'that'-clauses is needed to defeat the argument Graham Priest gave in his paper 'The Logic of Paradox' in support of his Dialetheism. But that attention also leads to a realization of the place and prevalence of indexicality in natural language. Let us just review the chain of thought that connects 'that'-clauses so intimately with indexicality, since it will take us much further.

Remember that Priest said: 'According to Gödel's incompleteness theorem in any [appropriate] formal system there will be sentences that are neither provable nor refutable. But this is not all there is. For some of these unprovable sentences can be *shown* to be true, i.e., proved in the naive sense.' The error here, of course, occurs when Priest says 'some of these unprovable sentences can be *shown* to be true, i.e. proved in the naive sense'. For what is 'provable in the nave sense' are propositions, not sentences, as we have seen. Priest's 'provability in the naive sense' is the provability of propositions, while what he, with many others, would call 'the proof of a formula' is, instead, the derivability of that formula from certain others using some purely syntactic rules.

It was through more generally ignoring this fact that the sentential tradition got into its difficulty with the Liar Paradox, as we also saw. For while we can easily construct a sentence J such that J = 'sentence J is not true', i.e. '¬TJ', if we use Horwich's Equivalence Scheme instead of Tarski's Truth Scheme, then we do not get a contradiction. We can say that ¬TJ, since sentences themselves do not have any truth-value. But then ¬TJ ≡ ¬T'¬TJ', and not ¬T< ¬TJ> as would be needed to engender a paradox. Certainly if '¬TJ' = < ¬TJ> then there would be a contradiction, but that would involve equating a mentioned sentence with a 'that'-clause, which is just the error in Priest's argument above. Of course, again, if the sentence '¬TJ' was *unambiguous*, then the proposition < ¬TJ> would be the only thing that could be stated when using the sentence, and the truth of the proposition could be reckoned to be a fact about the sentence, leading us back from Horwich's Equivalence Scheme to Tarski's Truth Scheme. But *then* there would be a paradox with sentence J* = 'the proposition stated by J* is not true'. For then, by the Equivalence Scheme, <the proposition stated by J* is not true> is true if and only if the proposition stated by J* is not true, while the proposition stated by J* is <the proposition stated by J* is not true>. So all this means that '¬TJ' *cannot* be unambiguous, and attention *must* be given to indexicality in this region.

Anyone who thinks otherwise is forgetting the possibility that, outside of a context, no definite proposition is made in such cases, as no proposition would be made if the referring phrase ostensibly referring to a proposition or statement were replaced by a demonstrative such as 'this'. For then it would follow that the sentence on its own would not state anything definite to be not true. In fact it follows by *Reductio* from the supposed contradiction above, that the definite description 'the proposition stated by J*' must be non-attributive, i.e. Millian. The mental block people have had at this point thus most likely arises through a need, or hope for 'central control' like that which arose with eternal sentences: in a 'perfect language' descriptions used in referential phrases surely would be truly attributable to the objects referred to.

In understanding this in the case of names, our initial discussion of 'John is happy' is crucial, since it shows that the seemingly expected univocality of ordinary, everyday referential sentences does not hold. There are many people called 'John', and so it can only be a *chosen one of them* that is spoken about if the remark is referential — εx(x is called 'John'). It is for this reason that the 'sentence J' in 'sentence J is not true' need not have that same sentence as its reference. The *very same sentence* might have referred to a quite different sentence, under a different naming system, and so made a different proposition.

5

But can anything similar be said about the mainline tradition's inattention to indexicality in connection with other paradoxes that have be-devilled twentieth century Formal Logic? Here we get to the wider significance of the move from recognising 'that'-clauses, to realising the necessity of indexicality. For very much the same can in fact be said about the mainline tradition's inattention to indexicality with respect to paradoxes such as the Paradox of Heterologicality, the Paradox of Predication, and Russell's Paradox. The predicates 'is not applicable to itself', 'is a property it does not possess' and 'is not a member of itself', involved in these paradoxes, each contain a context sensitive pronoun, whose referent therefore cannot be captured in a context-independent language.

A predicate (in the old, and, outside of Formal Logic books, still current sense) is a proper part of a sentence: it is that part of a sentence that remains after the subject is removed. Thus commonly, in English, the predicate is the latter part of a sentence, the part that follows the subject, that commonly comes first. In this way the predicate in 'x is not a member of x' is 'is not a member of x', and the subject is the 'x' that has then been removed. On the other hand the form of the whole sentence is '(1) is not a member of (1)', and this has been thought of as a kind of 'predicate', following Frege. On this variant understanding of 'predicate' there is also a different understanding of 'subject'. A subject in this alternative sense is not what is maybe at the start of a sentence, but becomes a term or expression that may recur throughout the sentence. Thus if '(1) is not a member of (1)' is taken as the 'predicate' in 'x is not a member of x', then 'x' becomes the 'subject' in this second sense, because it replaces '(1)' at all occurrences, not just at the start.

The distinction enables us to see that something different is said of A and of B when, for example, we say of each that he shaves himself. For what is then predicated of each does not have the verbal form '(1) shaves (1)', but simply 'shaves himself', and the 'himself' has a variable referent, dependent on its contextual antecedent. So different properties are attributed to A and to B: the property of shaving A in the one case, and the property of shaving B in the other. Of course, all those who shave themselves might still contingently (and so when only a finite set was involved) share a further property, and form a set of those who have that property, as when, for instance, they are all together in a room: $(x)(Rxx \equiv Px)$. But there is no necessity that there is such a 'P' for all 'R', i.e. there is no logical equivalent of 'Rxx' of the form 'Px' in general.

To see why one must first remember that there is a grammatical difference between equivalence and identity, especially when it is said that predicates are functions. For it has been common since Frege's time to take such equivalences as 'Rab $\equiv$ T' and 'Rab $\equiv$ F', where 'T' and 'F' are a tautology and a contradiction respectively, to be analogous to functional identities like 'g(a, b) = 1' and 'g(a, b) = 0' where '1' and '0' are names of truth values. But equivalences are formed between used sentences, while identities are between referring phrases, and (despite Frege) used sentences are not referring phrases, only their nominalisations are. Thus John's being a bachelor is identical with his being an unmarried male; but John is a bachelor if and only if he is an unmarried male. So equivalences like 'Rab $\equiv$ T' are not identities like 'g(a, b) = 1', and indeed the equivalences above are equivalent to things of quite a different character, namely just 'Rab' and '¬Rab' respectively.

More significant still is the fact that the two equivalences above are by their nature contradictories, while the two identities above are only contraries. So the pair of identities cannot match the pair of equivalences even in terms of the relation between them. And that leads us to the crux of the matter. For while it is certainly possible to define a function f(x) such that, say

Rxx $\equiv$ (f(x) = 1),

¬Rxx $\equiv$ (f(x) = 0),

that not only does not make the relation the function, also the right hand sides of these two equivalences here cannot be figured as involving the same predicate of x. If 'f(x) = 0' was replaced by 'f(x) ≠ 1', then there would be the same predicate of 'x' on the right hand sides, in one place negated; but no specific function would then be defined. It follows that the widely held view that binary relations where the argument is repeated are equivalent to one-place predicates cannot hold in general, i.e. it is not a logical truth that

(R)(∃P)(x)(Rxx ≡ Px).

The point resolves a number of puzzles that have bedevilled twentieth century logic. In connection with Grelling's Paradox, for instance, a problem arises when we use such a word as 'heterological' for what 'x' is when 'x' does not apply to 'x'. For then the variable within the (old-style) predicate 'does not apply to "x"' is obscured, since such words are properly used only for constant predicates. If instead we use 'not self applicable', the variable nature of the predicate is more apparent, although we still might forget that substituting 'not self applicable' for 'x' in:

'x' is not self applicable if and only if 'x' is not x,

means substituting it for 'self' as well as 'x', since there are four references to 'x' in the statement, and not just three. Substituting 'not self applicable' ('NSA') for 'x' in this statement does not lead to the contradictory

'NSA' is NSA if and only if 'NSA' is not NSA,

but to

'NSA' is not 'NSA' applicable if and only if 'NSA' is not NSA,

which is unexceptionable.

Likewise in Quine's 'yields a falsehood when appended to its own quotation': here 'its own' has a reference dependent on the subject to which the predicate is attached, and so the predicate is only properly expressible in a context sensitive language. In " 'yields a falsehood when appended to its own quotation' yields a falsehood when appended to its own quotation" the second 'its own' has a reference determinable from the context, but the first 'its own' has no determinate reference at all, since it has been abstracted from any context by merely being part of a mentioned expression. As a result, in the required full sense, that mentioned expression, without a determinate reference for the pronoun, can never get appended to *its own* quotation, since when appended it is not mentioned but used, and thereby a portion of it necessarily has a different sense from when it is not appended, through the context sensitivity which gives a determinate reference to the pronoun in it.

The difficulty with seeing all of the above compounds if one is attached to Frege's 'Px' symbolisation of elementary sentences. If 'P' there stands for a property or concept, and one abstracts 'x' from 'Px', then what is left but a referring phrase to that property or concept? This form of symbolisation arose because of Frege's assimilation of concepts to functions. Concepts were not objects, he believed, making 'the concept *horse*', which Frege agreed was a referring phrase to an object, paradoxically not a referring phrase to a concept. Instead, concepts, like 'P', were functions, and with the subject x taken as the argument of the function, the value of the whole subject-predicate expression was supposedly a truth value, in parallel with 'Px = T'. But this construal of truth as the value of a function is erroneous, as we have just seen.

More directly, if we wrote for the form of an elementary predication, not 'Px', but 'x has P', where 'P' is a referring phrase to the property or concept, it would be more clear that abstracting the subject leaves not a referring phrase to that property or concept, but merely *the predicate* in question, 'has P'. In fact, in the symbolism 'Px' the 'P' is doing double duty for a predicative expression involving a finite form of the verb, like 'is a horse', along with a referring phrase to the property alone, like 'being a horse', which would allow objectival second-order quantification. But, of course, on the latter reading of 'Px', 'x, the property P' is not a coherent sentence, giving rise to the problem of how a proposition is unified, and is not just a list of names. It characterises the 'bracketed' approach to Fregean Logic which is more prevalent, that writers on 'The Unity of the Proposition', and Frege's problem with the concept *horse*, do not commonly remember in their own practise the difference between 'is a horse', and its nominalisation 'being a horse', so that the absence of predicate nominalisations in Frege's symbolism is not brought up as an evident difficulty with it (c.f. Slater 2000). There is no lack of unity in 'Dobbin is a horse' like there is in 'Dobbin, being a horse', and the predicate in the former clearly does not refer to any concept, since it is not a referring expression at all, not being nominalised, as in the latter. It is a predicate like 'is a horse' which is functional, needing a subject term to be inserted to form a complete thought, and that leaves the associated concept phrase 'being a horse' to be unproblematically the designator of an (abstract) object.

That, of course, relieves us of one difficulty with the Paradox of Predication: '$(\exists P)(x = P \ \& \ \neg Px)$' is ungrammatical. The objectival quantification '$(\exists P)(x = P \ \& \ \neg(x \text{ has } P))$', in which 'P' is a referring phrase to a property, is not ungrammatical, by contrast. Alternatively (c.f. Kneale and Kneale 1962, 602), if one introduces the nominaliser '§', so that '§yPy' refers to the property while 'P' remains descriptive (and so equivalent to 'has '§yPy'), that would allow a further expression, '$(\exists P)(x = \S yPy \ \& \ \neg Px)$', to be well formed, in which the quantification is substitutional, with 'P' then an open predicate. In either case, however, we would have the question, as with Heterologicality, and Quine's Paradox, whether the resulting predicate expressed a single, unambiguous property. Indeed, if

$$(\exists P)(x = \S yPy \ \& \ \neg Px) \equiv Qx,$$

then there would be a contradiction when $x = \S yQy$. That contradiction proves indirectly that there is no single property, but what is the direct proof of this? How is it that '$(\exists P)(a = P \ \& \ \neg(a \text{ has } P))$', and '$(\exists P)(b = P \ \& \ \neg(b \text{ has } P))$', do not say a and b have the same property? The matter is now plain. Certainly these may seem to say the same thing about a as about b, but the sameness is only in the linguistic expression, which is systematically ambiguous. The common predicate 'is a property which it does not possess' contains a pronoun which refers back to the given subject term, so while the same syntactic item is involved, it is context sensitive with respect to its referent, which means that the whole is replaceable by 'is a property which a does not possess' and 'is a property which b does not possess', respectively, in the two cases. Geach once considered this kind of possibility, in connection with Heterologicality, but he thought one could remove the variability in other re-formulations (Geach 1972, 90). Certainly one can move the context sensitive element in 'is a property which it does not possess' to another place, as in 'is a property, but does not possess that property', or 'is, but does not possess the same property', but the referent of 'that property' and 'the same property' is variable as before.

As we shall see, similar points hold with respect to other paradoxes of much the same structure. In all, therefore, there has been 'a cloud of philosophy' which is now 'condensed into a drop of grammar', to quote Wittgenstein.

CHAPTER 10

QUINE'S OTHER WAY OUT

1

It is shown that, on the traditional, grammatical notion of a predicate as the remainder of a sentence once the subject term has been removed, there is no problem with Russell's Paradox, or comparable paradoxes such as Grelling's, and the Paradox of Predication. The standard formal ban on substituting predicates involving free variables into schemas where those variables would become bound is enough to prevent the standard paradoxes from developing. The re-arrangements required in the foundations of Set Theory to incorporate this insight are then discussed, and the consequences for the closely related matters Diagonalisation, and Cantor's Theorem explained.

2

I have pointed out in several places before (Slater 2004b, 2005b, and Appendix 2) that the Fregean tradition mixed up predicates with the forms of sentences. A predicate (in the old, and, outside of Logic books, still current sense) is a proper part of a sentence: it is that part of a sentence that remains after the subject is removed. Thus commonly, in English, the predicate is the latter part of a sentence, the part that follows the subject that commonly comes first. In this way the predicate in 'x is not a member of x' is 'is not a member of x', and the subject is the 'x' that has then been removed. On the other hand the form of the whole sentence is '(1) is not a member of (1)', and this has been thought of as a kind of 'predicate', following Frege. On this variant understanding of 'predicate' there is also a different understanding of 'subject'. A subject in this alternative sense is not what is maybe at the start of a sentence, but becomes a term or expression that may recur throughout the sentence. Thus if '(1) is not a member of (1)' is taken as the 'predicate' in 'x is not a member of x', then 'x' becomes the 'subject' in this second sense, because it replaces '(1)' at all occurrences, not just at the start.

The distinction enables us to see that something different is said of a and of b when, for example, we say of each that he shaves himself. For what is then predicated of each does not have the verbal form '(1) shaves (1)', but simply 'shaves himself', and the 'himself' has a variable referent, dependent on its contextual antecedent. So different properties are attributed to a and to b: the property of shaving a in the one case, and the property of shaving b in the other. Of course, all those who shave themselves might still contingently share a further property, and so form a set of those who have that property, and who, incidentally, are all of those who shave themselves, as when they are all together in a room: $(x)(Rx \equiv Sxx)$. But there is no necessity that there is such an 'R' for all 'S', i.e. there is no logical equivalent of 'Sxx' of the form 'Rx' in general.

The point resolves a number of puzzles that have bedevilled twentieth century logic. For, in connection with Grelling's Paradox, a problem arises when we use such a word as 'heterological' for what 'x' is when 'x' does not apply to 'x' (see Slater 1973). For then the variable within the (old-style) predicate 'does not apply to 'x'' is obscured, since such words are properly used only for constant predicates. If instead we use 'not self applicable', the variable nature of the predicate is more apparent, although we still might forget that substituting 'not self applicable' for 'x' in:

'x' is not self applicable if and only if 'x' is not x,

means substituting it for 'self' as well as 'x', since there are four references to 'x' in the statement, and not just three. Substituting 'not self applicable' ('NSA') for 'x' in this statement does not lead to

'NSA' is NSA if and only if 'NSA' is not NSA,

but to

'NSA' is not 'NSA' applicable if and only if 'NSA' is not NSA,

which is unexceptionable.

The same applies to Russell's Paradox, the Paradox of Predication, and other forms of Grelling's Paradox. For, notoriously, if we try to represent 'x is not a member of itself' as 'x is a member of R' for some fixed 'R', then a contradiction ensues. But none does if we respect the variable nature of 'itself'. What x is necessarily a member of, for instance, if it is not a member of itself, is its complement. But 'its complement' contains the contextual element 'its', and so in

x is a member of its complement if and only if x is not a member of x,

(i.e. $x \in x' \equiv x \notin x$) substitution of 'its complement' ('IC') for 'x' leads not to the contradictory

IC is a member of IC if and only if IC is not a member of IC,

(i.e. $x' \in x' \equiv x' \notin x'$) but to the unexceptionable

IC is a member of IC's complement if and only if IC is not a member of IC,

(i.e. $x' \in x \equiv x' \notin x'$) once one remembers that there is a variable item in the predicate 'is a member of its complement'.

In the Paradox of Predication the concern is with 'x is a property it does not possess', or 'x is a property but does not possess that property', i.e.

$(\exists P)(x = P \,\&\, \neg(x \text{ has } P))$.

But this is

$x = P^* \,\&\, \neg(x \text{ has } P^*)$,

with

$P^* = \epsilon P(x = P \,\&\, \neg(x \text{ has } P))$

in the epsilon reduction, which clearly shows that the property attributed to x in the (old-style) predicate is not constant, but varies with x. Likewise with Grelling's Paradox in the form ''x' does not possess the property it expresses', or ''x' expresses but does not possess a certain property' i.e.

$(\exists P)('x' \text{ expresses } P \,\&\, \neg('x' \text{ has } P))$.

3

However, it has recently come to my attention that there is another way of obtaining this conclusion using a standard feature of formal logic (c.f. Slater 2010c). For if the substituted 'F' in the naive abstraction scheme

$(\exists y)(x)(x$ is a member of $y \equiv Fx)$,

had to be a predicate in the old style, then the substitution of 'is not a member of x' for 'F' would violate a formal restriction. If one tried to derive Russell's Paradox from this abstraction scheme by substituting the predicate 'is not a member of x' for 'F', to get 'x is not a member of x' for 'Fx', then this would violate the restriction that variables free in a predicate must not be such as to be captured by quantifiers in the scheme into which the predicate is substituted (c.f. Quine 1959, 141). For the variable 'x' in 'is not a member of x' would become bound by the quantifier '(x)'.

There is no problem with introducing occurrences of other variables in the substituted predicate, but there is a quite general problem with bringing in a variable free in the substituted predicate that would be bound in the scheme it is substituted into. In an example from Quine, consider the substitution of 'Gx' for 'F' in

$Fy \supset (\exists x)Fx.$

This implication is formally valid, so the given substitution is improper since it would yield

$Gxy \supset (\exists x)Gxx,$

which is invalid (c.f. Quine 1959, 144).

Quine himself overlooked the way this point provides a way out from Russell's Paradox. That was no doubt because the novel Fregean grammar was burnt well into him. In the way Fregeans think of it, it is quite proper that, in the scheme of naive abstraction, 'F(1)' be replaced by '(1) is not a member of (1)', to yield

$(\exists y)(x)(x$ is a member of $y \equiv x$ is not a member of $x)$.

Putting it this way, one is using Quine's device of 'placeholders' to indicate the argument-places of 'F(1)'. The point to note is that the complex 'predicate' (strictly 'form of a sentence') that then replaces 'F(1)' does not contain any occurrences of 'x', hence the above bar on capturing seemingly does not apply. Fregeans would think of themselves as substituting '(1) is not a member of (1)' not for 'Fx' but for 'F(1)', where the argument places marked by '(1)' are filled by whatever fills the argument place of 'F(1)' — in the above case 'x'.

But if we keep to the traditional notion of predicate as the remainder of a sentence after the removal of (in English) the first occurrence of its subject, then clearly Quine's restriction will enable us to escape the paradox that results from the Fregean way of looking at the matter. More exactly, it will enable us to escape from paradox with any substitution into the abstraction scheme

$(\exists y)(x)(x$ is a member of $y \equiv Fx)$,

that does not violate the above bar on capturing. For the further point that needs to be made is that that does not preclude having further abstraction schemes applying when there is reflexivity in the predicate. There is no problem with replacing the 'F' above with any constant, old style predicate, or even such a predicate involving another variable, like 'Rz'. But being unable to replace the above 'F' with 'Rx' leaves us with the need for an abstraction scheme applicable when 'Rxx' is on the right hand side. That is no problem, however, since the way to handle relations quite generally, and so equally when the subject is repeated, is to bring in sets of ordered pairs.

If a is shaving a, then, as before, a has the property of shaving a (also the property of being shaved by a). But a also stands in a relation to himself: he and himself form a shaving (i.e. shaver-shaved) pair. Moreover, if a is shaving a, and b is shaving b, then the same relation is involved — a relation that a holds with himself, and b holds with *himself* (sic, notice the change of referent); and that relation is not a specifically 'reflexive relation', since it is the same relation that a would have with b, if a were shaving b. Thus quite generally,

$(\exists y)(x)(z)(<z,x>$ is a member of $y \equiv Rzx),$

and the same y is involved if z=x, even though y is not just a set of ordered pairs whose members, in each case, are the same. But, in the particular case

$(\exists y)(x)(<x,x>$ is a member of $y \equiv x$ is not a member of $x),$

we only get, on substitution,

$<y,y>$ is a member of $y \equiv y$ is not a member of $y,$

which is not a contradiction.

Surprisingly, therefore, we must conclude that, from a traditional perspective, Frege got into his problem with Russell's Paradox through forgetting the applicability of the elementary notion of 'being free for' to the case.

4

Of course the elementary, and rather banal principles above are the basis for more interesting and complicated results, since once sets for elementary predicates are defined, those for non-elementary predicates can be constructed out of them by standard set-theoretic processes.

Thus, for a start, from the Abstraction Axiom (given Extensionality, which ensures uniqueness of referent for the epsilon term) we can invariably write '$\{x: Px\}$' or '$\epsilon z(y)(y \in z \equiv Py)$' for the set of Ps (where the (old-style predicate) 'P' in 'Py' contains no occurrence of 'y'). Repeated variables in a relation must be handled differently, as above, but since

'x is P and x is Q'

is the same as

'x is P and Q',

and

'x is R to y, and x is S to y'

is the same as

'x is R and S to y',

etc., the repeated variables in conjunctions like 'Px & Qx' can be handled using the normal definition of set intersection. Thus:

$\{x: Px \,\&\, Qx\} = \{x: x \in (\{y: Py\} \cap \{y: Qy\})\}.$

Likewise with the union of two sets, and the complement of a set:

$\{x: Px \vee Qx\} = \{x: x \in (\{y: Py\} \cup \{y: Qy\})\},$
$\{x: \neg Px\} = \{x: x \notin \{y: Py\}\}.$

The null set can then be defined as the intersection of $\{x: Fx\}$ and $\{x: \neg Fx\}$ (for any 'F'), i.e.,

$\varnothing = \epsilon y(x)(x \in y \equiv Fx) \cap \epsilon y(x)(x \in y \equiv \neg Fx),$

and the universal set likewise as the union of $\{x: Fx\}$ and $\{x: \neg Fx\}$ (for any 'F').

As for the standard axioms of Set Theory, the present approach has the advantage of making most of them redundant. Thus the Axiom of Regularity is not required since there is nothing suspect about expressions like '$x \in x$', and their more complex kin. Some of the functions of the Axiom of Choice are taken over by the properties of epsilon terms (as in Bernays' formulation of Set Theory, see Bernays 1968). For

$(\exists x)(x \in y) \equiv \epsilon x(x \in y) \in y,$

and so the appropriate epsilon term always provides a selection from a non-empty set. The Power Set Axiom follows using Abstraction on the definition of a subset, for given

$y \subset x \equiv (z)(z \in y \supset z \in x),$

and

$(\exists z)(t)(t \in z \equiv t \subset x),$

then one can always define

$\Psi x = \{y: y \subset x\} \ (= \epsilon z(t)(t \in z \equiv t \subset x)).$

The Axiom of Pairs is now an immediate inference from Abstraction and the definition of the union of two sets, since

$(\exists y)(x)(x \in y \equiv x = z),$

yields

$x = z \equiv x \in \{t: t = z\},$

and so it follows, using the process of (finite) set union above, that

$(\exists y)(x)(x \in y \equiv (x = z \lor x = t)).$

The Axiom of Separation in the form

$(\exists y)(x)(x \in y \equiv (x \in z \ \& \ Px)),$

(where 'Px' is as before) is now an immediate inference from Abstraction and the definition of the intersection of two sets. But the Axiom of Separation is standardly expressed using, in place of 'Px', a formula in which 'x' might occur any number of times. So that will not follow in the present case, without a series of further assumptions like

$(y)(\exists x)(t)(t \in x \equiv <t,t> \in y).$

This (and its kin with larger ordered sets) clearly holds if the set corresponding to 'y' is finite, and so can be listed and not just known descriptively, i.e. 'intensionally'. But it cannot hold in general, since it is just this kind of assumption, we now see, that generates Russell's Paradox.

A significant departure from normal Set Theory immediately arises, however, because of the now marked difference between listable (and so finite) and unlistable (because potentially infinite) set intersections and set unions. For while we may have a set of individuals in the following case:

$(\exists t)(x)(x \in t \equiv (Rxa \lor Rxb)),$

for instance, we can only have a set of pairs of individuals in such a case as:

$(\exists t)(x)(<x,x> \in t \equiv (\exists y)Rxy).$

This is because

$(\exists y)Rxy$

is equivalent to

RxεyRxy,

i.e. to

x Rs what it Rs,

in which 'x' is repeated. The existential quantifier brings in a pronominal phrase 'what it Rs' ('εyRxy') which, like the 'himself' in 'x shaves himself' has a variable referent.

So care must also be taken with, for instance, such equivalences as

$Rss \equiv (\exists t)(s = t \;\&\; Rtt)$.

The R.H.S. here looks like it might be of the required constant form 'Ps', and so the further assumption above may seem to be automatically satisfied. Thus 's shaves himself' is equivalent to 's is someone who shaves himself' and the predicate 'is someone who shaves himself' might seem to have a constant sense. The subject-predicate structure of the R.H.S., however, is more fully displayed in its epsilon equivalent:

$s = t^* \;\&\; Rt^*t^*$,

where $t^* = \epsilon t(s = t \;\&\; Rtt)$. So in the old-style predicate in question (i.e. the portion of this last expression after the initial 's') there are again further occurrences of the subject, making the referent of the pronoun 'someone' in 'is someone who shaves himself' not constant, but a function of the subject the predicate is applied to. So while there is a constant syntactic predicate, the epsilon analysis reveals it expresses a variable property, as again with 'shaves himself', 'does not apply to itself' etc.

What the above equivalence does logically ensure is that something of the following form is provable:

$(t)(<t,t> \in x \equiv <t,t,t,t> \in y)$.

But with 'Rss' as 's shaves himself' then, as before, only contingently (and thus only with a finite set being involved) could there be a z such that

$(t)(t \in z \equiv <t,t> \in x)$.

So 'Separation' in its traditional form is not automatically guaranteed, and that also means that the Axiom of Choice, on which the full form of Separation is clearly based must not be assumed in general. For one moves in the further assumptions above from a set of ordered sets with iterated members to a set that selects just one member from each of those ordered sets (the problem being not in the selection of the various members, but in whether there is a set of all those selected when it would have to be given descriptively, or 'intensionally', being infinite). That leaves Abstraction, and Extensionality as the only two set-theoretic principles that are totally justified.

5

There are consequences for the understanding of Diagonalisation, of course, with which Russell's Paradox is closely related. For what has been called 'Cantor's Theorem' seems to show that the power set of any set has greater cardinality than the set itself. If 'x' ranges over members of a set, and 'Sx' over correlated subsets of the set, then Cantor argued that

$(x)(x \in S_y \equiv x \notin S_x)$,

must define a further subset, S_y, i.e. the 'y' cannot name a member of the set. But if that was so then it would follow that there could be no universal set.

For each of its subsets would have to be a member of it, being sets, making the cardinality of the universal set at least as great as the cardinality of its power set — a contradiction. But a universal set is easily defined, as before. So there must be something wrong with Cantor's argument, and what is wrong is now easy to diagnose. For what is true, for a start, is merely that

$$(\exists z)(x)(<x,x> \in z \equiv x \notin S_x),$$

so Cantor needed a further premise

$$(\exists y)(t)(t \in y \equiv <t,t> \in \epsilon z(x)(<x,x> \in z \equiv x \notin S_x)),$$

to establish that his 'theorem' held in general.

Likewise with other forms of 'Cantor's Theorem'. For given a defined sequence of functions of one variable, $f_x(y)$, onto $(0,1)$ then

$$F_r(x) = 1 - f_x(x)$$

will define a different function, i.e. the 'r' will not be one of the 'x's. But in the extreme case, where the sequence contains *all* functions of one variable onto $(0,1)$, evidently no new function can be defined in this way, without contradiction. So it is not just that there might be something like a 'non-recursive function' in such a case beyond recursive ones. What there is in the extreme case is a sequence of functions without any definable function of one variable generating $f_x(x)$, because from the index 'x' there is no definable function generating the function with that index, and so no 'F' such that $f_x(x) = F(x)$.

The situation, in other words (c.f. Slater 2000b, 94-5), parallels that for computable functions. For while all computable functions of one numerical variable onto $(0,1)$ are enumerable, there is no way to specifically enumerate just those that have completely defined values (i.e. which are not just partial but total functions), otherwise the halting problem would be solved. Hence the ordinal numbers of those functions that are total, although denumerable, are not enumerable. There is, in other words, a further kind of expression, which is like that for a binary 'decimal' except certain places are undefined. These expressions are enumerable, but diagonalisation does not produce a further one of them, since neither $f_m(n)$, nor $1 - f_n(n)$ need equal anything. Amongst the functions which generate these expressions are all the total functions of one variable, but we cannot, in general, determine which these functions are. Even if $f_m(n)$ is total, which function it is is only determinable from its ordinal place amongst all the computable functions of one variable, not from its ordinal place amongst the total functions of this sort, with the result that, if the latter is 'm', then $f_m(n)$ is not a calculable function of m.

Of course, if one *specifies* a sequence just of total functions that makes it the case that which function is the mth in that sequence is determinable from m, and $1 - f_n(n)$ will then be a further, distinct total function of n. But it is only the specification of such a sequence which makes $f_m(n)$ a function both of m and of n, and so there is no further diagonal function in an unspecified case, much as there was no diagonal set in the extreme case before.

CHAPTER 11

LOGIC IS NOT MATHEMATICAL

1

This chapter summarises some of the main points that have been made in this book, before extending them further. I first show how twentieth century Set Theory got into its greatest tangle through, amongst other things, regarding relational remarks like 'Rxy' as binary functions. I then show how the lack of indexicality, and 'that'-clauses in the supposedly 'perfect languages' of Modern Logic led that subject into its intractable difficulties with the Theory of Truth. Both errors arose not only through a contempt for ordinary language, but also through the related failure to recognise that being logical is not a matter of being brainy, but of being coherent. It is not a mathematical talent, but a literary one. Later in the chapter I go on to demonstrate this same conclusion with respect to Modal Logic and General Intensional Logic, and in particular with respect to fictions, since these are the central items that have been misunderstood, as is witnessed in some recent writings of Graham Priest.

2

I start with some incidental matters that merely illustrate the consequences of inattention to ordinary language. Thus the paradox about 'heterologicality', in parallel with Russell's Paradox, involves a mistake about reflexive pronouns that has been prevalent in formal logic since Frege's times. For it is common to stipulate, for instance, that a word be called heterological just so long as that word is not applicable to itself. But if we use a word such as 'heterological' for what 'x' is when 'x' does not apply to 'x', then the variable within this predicate is obscured, for such words are properly used only for specific predicates. If instead we use 'not self applicable', the variable nature of the predicate is more apparent, although we still might forget that substituting 'not self applicable' for 'x' in:

'x' is not self applicable if and only if 'x' is not x,

means substituting it for 'self' as well as 'x', since there are four references to 'x' in the statement, and not just three. Substituting 'not self applicable' ('NSA') for 'x' in this statement does not lead to the contradictory

'NSA' is NSA if and only if 'NSA' is not NSA,

but to the unexceptionable

'NSA' is not 'NSA' applicable if and only if 'NSA' is not NSA.

Why people were led into this mistake was because of a shift in the notion of a predicate that Frege also was responsible for. For it has been common since Frege's time to take such equivalences as 'Rab ≡ T' and 'Rab ≡ F', where 'T' and 'F' are a tautology and a contradiction respectively, to be analogous to

functional identities like 'g(a, b) = 1' and 'g(a, b) = 0' where '1' and '0' are
names of truth values. So 'Rxx' seems to be 'Px' for some 'P'. But equivalences
are between used sentences, while identities are between referring phrases, and
(despite Frege) used sentences are not referring phrases, only their nominalisations are. So equivalences like 'Rab ≡ T' are not identities like 'g(a, b) = 1',
and indeed the equivalences above are equivalent to things of quite a different
character, namely just 'Rab' and '¬Rab' respectively. More significant still is
the fact that the two equivalences above are by their nature contradictories,
while the two identities above are only contraries. And that leads us to see
that while it is certainly possible to define a function f(x) such that, say

$$Rxx \equiv (f(x) = 1), \neg Rxx \equiv (f(x) = 0),$$

that not only does not make the relation the function, also the right hand sides
of these two equivalences here cannot be figured as involving the same predicate
of x. If 'f(x) = 0' was replaced by 'f(x) ≠ 1', then there would be the same
predicate of 'x' on the right hand sides, in one place negated. But no specific
function would then be defined. It follows that the widely held view that binary
relations where the argument is repeated are equivalent to one-place predicates
cannot hold in general, that is, it is not a logical truth that

$$(R)(\exists P)(x)(Rxx \equiv Px).$$

A predicate (in the old, and, outside of Logic books, still current sense) is a
proper part of a sentence: it is that part of a sentence that remains after the
subject is removed. Thus commonly, in English, the predicate is the latter part
of a sentence, the part that follows the subject that commonly comes first. In
this way the predicate in 'x is not a member of x' is 'is not a member of x', and
the subject is the 'x' that has then been removed. On the other hand the form
of the whole sentence is '(1) is not a member of (1)', and this has been thought
of as a kind of 'predicate', following Frege. On this variant understanding of
'predicate' there is also a different understanding of 'subject'. A subject in this
alternative sense is not what is maybe at the start of a sentence, but becomes
a term or expression that may recur throughout the sentence. Thus if '(1) is
not a member of (1)' is taken as the 'predicate' in 'x is not a member of x',
then 'x' becomes the 'subject' in this second sense, because it replaces '(1)' at
all occurrences, and not just at the start.

One of the things that obscures the previous point, for logicians, is that in the
supposedly 'perfect languages' of modern logic there is no passive transform.
In natural language one can transform the active form 'a shaves b' into an
equivalent passive form 'b is shaved by a'. In so doing one changes the subject
from 'a' to 'b', and the predicate from 'shaves b' to 'is shaved by a'. But in the
language of modern predicate logic both sentences could be written as having
the form 'Rab', with 'R' being read appropriately, and neither of the 'a' or
'b' being specifically the subject. So, even though the distinction between a
relation and its converse is retained, the old notion of a predicate gets lost, and
the modern notion replacing it, of the form of a sentence, takes over.

But another thing that obscures the point, highly relevant, now, to what
follows, is the absence of nominalisations in the supposedly 'perfect languages'
of modern logic. That hides the previously shown difference between relations
and functions. For there is no way to say, in the standard languages of modern

logic, 'That Jack is a bachelor is the same as his being an unmarried man', only 'Jack is a bachelor if and only if he is an unmarried man'. So the two get conflated, with the consequent distinction between equivalence and identity being muddied. One thing that distinguishes them is that the above identity only entails the above equivalence, and is not logically equivalent to it. For the equivalence is only a material one. As we shall see, the situation is not improved if a necessity operator is put before the material equivalence, since the identity of propositions is not given by logical equivalence.

But there is also no indexicality in the languages of modern logic. Quine, notably, thought that he could remove such matters by constructing what he called 'eternal sentences.' For instance, people might try to replace the indexical 'He is happy', said of John, with 'John is happy', to get a sentence with a fixed truth value, which might, amongst other things, fit into the Truth Scheme. But the tense in 'John is happy' makes this sentence indexical another way. So Quine only held that 'He is happy', said of John *at time t*, should be replaced with the *eternal sentence* 'John at time t is happy'. But Quine's transformation is impossible, in total, since the number of words in a language must be denumerable, while the number of times could be non-denumerable. Moreover we now have to deal with Modal Logic, and so deal with 'He is happy', said of John at time t in possible world w. Even more clearly this cannot be replaced with the eternal and transworld sentence 'John at time t in possible world w is happy'. So one cannot always find a sentence with a fixed truth value for the Truth Scheme to apply to, i.e. get 'central control' *within the language* of what is being said. For instance, can one even say: 'John at time t is happy' is true $\equiv$ John is happy at time t? Which person named 'John' is being spoken about? Clearly the formal tradition has been presuming that any name in a sentence refers unambiguously just to one thing. But language cannot work like that: names are themselves indexical. The point, note, also applies to Gödel numbers, since what sentence such a number refers to depends on the system of numbering used.

A recognition of the previous points about indexicality and sentence nominalisations gets one out of the Liar, and associated paradoxes. Tarski had well-known problems with the Liar Paradox, but he also had problems with indexicals, since he thought that his Truth Scheme still worked when they were in the sentences considered. Thus Carnap, in his 'Intellectual Biography' says: I wondered how it was possible to state the truth-condition for a simple sentence like 'this table is black'. Tarski replied, "This is simple: the sentence 'this table is black' is true if and only if this table is black". But the Truth Scheme " 'p' is true $\equiv$ p" does not apply when indexicals are involved, since one cannot say: 'He is happy' is true $\equiv$ he is happy. Even Horwich, who has studied the matter more than most, tries to keep indexical declarative sentences (along with all others exhibiting ambiguity and context sensitivity) within the purview of 'the disquotation schema'. He does so by inventing a device, which with 'this table is black', for example, would make 'this table', when merely quoted on the left hand side of the schema, have the same referent as when unquoted on the right hand side. But it is not the *mentioned sentence* 'this table is black' that is then said to be true, it is the *proposition* it makes at the place where it is *used*. It is not that 'this table is black' is true ≡ this table is black, but that *that this table is black is true* ≡ this table is black. Horwich has in this way been led to think there are closer parallels between sentences and propositions than in fact there are, and in particular he has been led to believe that propositions fall foul of Liar-type paradoxes, just like sentences. But as we have seen they clearly do not, once the relevant distinctions between sentences and propositions have been made entirely clear.

It is because people, under the influence of the current formal tradition forget the indexicality in language that they get into such problems as the Liar Paradox. For one frequently hears arguments about sentences like 'This is not true' generating contradictions — even 'true contradictions'. Surely, if that sentence is true then, because of what it says then it must be not true, and so not true, absolutely. But also, surely, then it is not not true, i.e. true, again because of what it says. The reasoning, of course, is fallacious because it hides the presumption that the verbal item 'this' in the sentence has a specific reference that the sentence alone does not carry. The intention, or attempted intention, is to refer, with 'this' not to the sentence itself but to some proposition it is taken to express given the meaning then associated with the demonstrative. But what proposition is that? Until the referent of the subject term is provided there is no proposition expressed, and yet a referent would be needed prior to the intended proposition being identified via what was predicated. The situation is no better if one gives, say, numbers to sentences — even Gödel numbers — since they too could have many referents. If sentence 1 is 'sentence 1 is not true' then the referent of the 'sentence 1' in quotes is still to be given, and must not be presumed to be the same as the referent of the 'sentence 1' outside of the quotes. The reference of a term like 'sentence 1' is dependent on the context of its use, and thus it is not confined to be the reference given to it in the present context. Because of this the disquotation required in Tarski's T-scheme is not guaranteed, invalidating it as a principle.

But also the Liar Paradox is now seen to arise only with Tarski's T-scheme. For Horwich's 'That p is true', while it is of the subject-predicate form, is equivalent to 'It is true that p', which is of an operator form. And 'It is true that p' involves the operator 'it is true that' which is the null or identity operator in the modal system KT, which is consistent.

3

In addition to the resultant consistency of propositional logic with the correct truth scheme, the now *propositional* truth scheme 'T<p> ≡ p' means, for instance, that the Principle of Bivalence (it is true that p or it is true that ¬p) is logically equivalent to the Law of the Excluded Middle (p v ¬p). Also lines in a truth table such as 'If it is true that p and it is not true that q, then it is not true that p & q' are equivalent to propositional theses like '(p & ¬q) ⊃ ¬(p & q)'. So the move to an intensional semantics not only produces a semantically closed language, but also collapses, amongst other things, the usual soundness and completeness proofs in propositional logic. A similar large-scale semantic collapse, with the removal of meta-languages, arises with variant expressions for modal terms. Indeed the change here is very extensive, and in quite novel directions. For instance, for a start, now 'that p is necessary' is *equivalent* to 'In all possible worlds that p is true', i.e. '(i)(V[p, i] = 1)' with 'i' ranging over all possible worlds, and the whole read as an operator expression, and so not with mentioned 'p' but used 'p'. Under Tarski's influence it seemed that one could compare 'It is necessary that p' with 'in all possible worlds 'p' is true', the latter form providing the 'semantics' of the former. But now we see that the two relevant expressions are in the same language, and are simply equivalent.

What of the possibility of variant Modal Logic's therefore? It has been common since the late 1950s, under the influence of some mathematical results of Kripke, to discriminate amongst modal systems, through the introduction of variant 'accessibility relations'. At the start, as in the first edition of Hughes and Cresswell's classic text, there was some attempt to give such relations an interpretation in terms of viewing one possible world from within another. Since that time, however, this matter has not been clarified, or indeed even pursued, and instead a pure mathematical fascination with the Kripkean technology seems to have taken over, with multiplication of quite abstract 'seeing' relations being generated. With the intensionalisation of necessity, however, we come to see there is no relativisation of the notion, since any conditionalisation of the number of possible worlds involved clearly produces something less than necessity. It remains, of course, that with respect to specific propositions one can discuss whether or not their necessity is itself necessary, or contingent, which has been one major division much discussed as a feature that defines modal systems. For while it may be the case that in all worlds it is true that p (i.e. (i)(V[p, i] = 1) with 'i' ranging over all possible worlds), that still leaves it open whether in all worlds it is true that it is necessary that p (i.e. whether (j)(V[(i)(V[p, i] = 1), j] = 1), with 'j' again ranging over all possible worlds.) For with 'V[p, i] = 1' being read as the operator expression 'it is true in world i that p', this becomes just another expression in the object language which may be iterated.

The point has a bearing on attempts that have been made, since Hintikka published his classic text *Knowledge and Belief* (Cornell 1962), to formalize these general intensional notions, along with others such as Obligation, in terms of possible worlds. It was hoped that this could be done with appropriate restrictions on the worlds involved. Thus it was said that a knows that p just so long as it is true that p in all the worlds compatible with what a knows. The problem of logical omniscience resulting from this definition has troubled many, however, since it follows that any necessary truth is known, and that everyone knows the logical consequences of any initial thing that they know. The consequences are equally accepted in AGM theories of Belief Revision. But while, of course, it remains that that p will be true in any world compatible with what a knows, if a knows that p, the reverse implication does not hold. In 'a knows that p' there is a relation between a and an hyperintensional object, so the relation is itself an hyperintensional one not a simple intensional one. Propositions are hyperintensional because their identity is given through translations, not logical equivalence. Logical equivalence as a criterion of identity would bring in irrelevant necessarily true conjuncts, for instance, and make all necessarily true propositions identical. The translational basis for the identity of propositions also defeats the common analysis of propositions in terms of sets of possible worlds. In fact the hard truth that has to be faced is that there is no *mathematical* analysis of propositions, only a *literary* one. The shift to a proper understanding of intensionality quite generally has this very radical dimension.

4

Specifically, in tune with the move to a literary point of view, we come to the detail of the analysis of individuals in intensional and hyperintensional constructions, and a fuller discussion thereby of fictions, and our relations with them. The logic of this is just an intensional epsilon calculus, and after recalling some of the background for that we can look at the issues in more human terms in connection with our involvement with books, films, plays and other theatrical phenomena. The basic point is that in an operator expression like 'V[p, i] = 1' any individual term within 'p' is being used, not mentioned, and so is in the voice of the speaker. That means it obeys the standard (extensional) identity laws, making intensional constructions transparent.

But we must first have completely clear what referential terms are. This, as we have seen, was obscured in Russell's Theory of Descriptions, and remains unclarified to this day because of the popularity of that theory, and the associated neglect of even the 'extensional' Epsilon Calculus. This name is something of a misnomer, however, since the use of epsilon terms already introduces intensional elements, because of their relation to choice.

By contrast, in Russell's seemingly entirely extensional theory there are, it will be remembered, three clauses with 'The King of France is bald'. These are 'there is a king of France', 'there is only one king of France' and 'he is bald'. Russell used an iota term to symbolise the definite description, but of course this is not an individual symbol since 'The King of France is bald' on Russell's account does not have the elementary form 'Bx'. Russell hypothesised that, in addition to the linguistic expressions gaining formalisations by means of his iota terms, there was another, quite distinct class of expressions, which would take the place of the variable in such forms as 'Bx'. He suggested that demonstratives might be in this class, but he could give no further formal expression to them. The epsilon theory of descriptions that settles the question was discussed in the first edition of Hughes and Cresswell's classic introductory text on Modal Logic. It originated with Routley, Meyer and Goddard, who, in their work on intensional contexts, made an explicit identification of definite descriptions with epsilon terms: The King of France = $\epsilon x(Kx \;\&\; (y)(Ky \supset y = x))$, (Goddard and Routley 1973, 558; Hughes and Cresswell 1968, 203).

We saw before which theorems in the epsilon calculus are behind this kind of identification. The standard epsilon calculus contains the axiom

$(\exists x)Fx \supset F\epsilon xFx$

(Leisenring 1969, Meyer Viol 1995), from which one can naturally obtain the equivalence between the two sides. This makes existence a matter of an individual 'living up to its name', i.e. actually having the properties inscribed in its referential description. There is then one theorem in particular which demonstrates strikingly the relation between Russell's attributive, and Donnellan's 'purely referential' understanding of referential terms. For, following on from the above equivalence it can be proved that

(1) $(\exists x)(Kx \;\&\; (y)(Ky \supset y = x) \;\&\; Bx)$,

is logically equivalent to

(2) $(\exists x)(Kx \;\&\; (y)(Ky \supset y = x)) \;\&\; Ba$,

where $a = \epsilon x(Kx \;\&\; (y)(Ky \supset y = x))$, this being the epsilon term arising from the first conjunct in (2). The first expression, as we have seen, encapsulates Russell's Theory of Descriptions, in connection with 'The K is B'; it involves the explicit assertion of the first two clauses, to do with the existence and uniqueness of a K. Since Donnellan, however (Donnellan 1966), we have realized that there are no preconditions on the introduction of 'the K' as an individual term. So 'The K is B', with 'The K' an individual term, may always be given a truth value, even if, sometimes, that truth value is merely an arbitrarily chosen one. 'Ba' properly formalises 'The K is B', since the cross-reference in (2) means that it reads 'There is a single K. It is B', and the descriptive replacement for the pronoun 'it', there, is 'The K'. On this basis (1) is better read 'A sole K exists and is B' rather than 'The K is B'. If the description in 'a' is non-attributive, i.e. if the first two clauses of Russell's account are not both true, and a sole K does not exist, then the referent of 'the K' is simply up to the speaker to nominate.

So we come to see that Russell's theory, when completed, turns out to be an intensional theory. For what epsilon terms formalise more generally are demonstratives, in line with Russell's identifying such as 'logically proper names', in his lectures on Logical Atomism. One major thinker familiar with the epsilon calculus who has missed this is Graham Priest, who took from Kneebone the idea that epsilon terms formalise indefinite descriptions, making the epsilon symbol replace 'an'. But if there is no uniqueness clause requiring a reading in terms of the definite descriptor 'the', then '$\epsilon x F x$' is best read 'that F'. This arises, for instance, when reading

'$(\exists x)Fx \;\&\; G\epsilon x F x$'

as

'There is an F, and that F is G'.

The epsilon terms in these kinds of contexts thus replace E-type pronouns in ordinary speech. E-type pronouns are anaphoric pronouns, and so are like demonstratives in that they point to other elements in the preceding discourse. Thus in this last case '$\epsilon x F x$' replaces 'It', while Routley, Meyer and Goddard's definite description 'the king of France' replaces 'He' in the context given before. In Donnellan's historic case with the phrase 'the man with martini in his glass' used referringly, the speakers are just selecting a referent for the epsilon expression '$\epsilon x(Mx \;\&\; Gx)$' when $\neg(\exists x)(Mx \;\&\; Gx)$, in line with the general semantics for epsilon terms. Thus if $(\exists x)Fx$ then the referent of '$\epsilon x F x$' is to be selected from amongst the Fs, while if $\neg(\exists x)Fx$ then it can be selected arbitrarily from the world at large.

How can something be the one and only K 'if there is no such thing', i.e. if there is nothing with the character inscribed in the term? That is where a second, and even more important theorem in the epsilon calculus is required:

$(Kc \;\&\; (y)(Ky \supset y = c)) \supset [c = \epsilon x(Kx \;\&\; (y)(Ky \supset y = x))]$.

For the singular thing is that this entailment cannot be reversed, so there is a difference between the left hand side and the right hand side, e.g. between something being alone king of France, and that thing being the one and only king of France. The difference is not available in Russell's logic, since only possession of the property can be formalised there. In fact Russell confused the two forms, since possession of an identifying property he formalised using the identity sign, viz. 'c = ιxKx', making it appear that some, maybe even all identities are contingent. But all proper identities are necessary, and it is merely associated identifying properties that are contingent. That means that in all possible worlds there is the same domain of discourse, but the individuals in that domain may change their properties, and even their individuating properties, from one world to the next.

What are the formal consequences of the above for modal, and general intensional logics? Clearly, if the same individual is involved, epsilon terms must be rigid across all worlds. Richard Routley presented several rigid intensional semantics, some objectival and some substitutional (Routley 1977, 185186). One of these semantics, for instance, simply took the first epsilon axiom to hold in any interpretation, and made the value of any epsilon term itself. Using such a rigid semantics, Routley, Meyer and Goddard obtained what has been called 'Routley's Formula', i.e.

L($\exists$x)Fx $\supset$ ($\exists$x)LFx,

by means of the following transformations. Routley's Formula holds for any propositional operator and any predicate, but they illustrated it in the case of necessity, and with 'Fx' as 'x numbers the planets'. Then, with 'ϵxFx' as 'the number of the planets', they said that from 'L($\exists$x)Fx', we can get 'LFϵxFx', by the epsilon definition of the existential quantifier, and so '($\exists$x)LFx', by existential generalisation over the rigid term (Routley, Meyer and Goddard 1974, 308; Hughes and Cresswell 1968, 197, 204).

We therefore see, amongst other things, not only that L($\exists$x)(x=a), because it is provable that something is a, if 'a' is an individual, i.e. epsilon term, but also that ($\exists$x)L(x=a), i.e. that the same thing is a in all possible worlds. In Kripkean semantics for Modal Logic, which was developed on the basis of Tarski's semantics, it seems we can discriminate between 'L($\exists$x)(x=a)' and '($\exists$x)L(x=a)' on the grounds that the former merely says that in every possible world there is something named 'a', while the latter says that the same thing is named 'a' in every possible world. But this way of looking at the formulae, we now see, involves a use-mention confusion. Certainly one can discriminate between 'in every possible world something is named 'a'', and 'the same thing is named 'a' in every possible world', but these are meta-linguistic, relational remarks, which ought to be symbolised

(y)($\exists$x)Nx'a'y,

and

($\exists$x)(y)Nx'a'y,

where 'Nx'a'y' says that x is named 'a' in world y.

One further consequence is that while individuals have eternal existence, they must be separated from any entities that merely have 'existence' in this world, or some other. For what, in connection with *individuals*, has 'existence' just in this world, or just in some other making them 'physical objects', and 'fictions', respectively are not the individuals themselves, but their *identifying properties*. To highlight the difference even more, we can say that *Aristotelian Realism* holds for such physical objects/fictions, whereas *Platonic Realism* holds for the associated individuals. The metaphysical point is important in connection with the shift from a mathematical to a literary point of view required for a proper understanding of intensions. For the difference between the two kinds of Realism is illustrated most clearly in the previous epsilon calculus theorem, which shows that

$(\exists x)(Kx \ \& \ (y)(Ky \supset y=x) \ \& \ Bx)$

(i.e. 'A sole king of France exists and is bald') is equivalent to

$(\exists x)(Kx \ \& \ (y)(Ky \supset y=x)) \ \& \ B\epsilon x(Kx \ \& \ (y)(Ky \supset y=x))$,

(i.e. 'A sole king of France exists. He is bald'). The first conjunct in the second expression is about certain identifying properties being instantiated. That is what must hold for a sole king of France to exist (contingently). The second conjunct there, however, is about a certain eternally existing individual — one that is a sole king of France if there is such a thing, but which still exists even if there is no such thing.

But not only through presenting Russell's formula as a conjunction do we enable a separation to be made between a true or false assertion about this world, namely the first conjunct delimiting existence and uniqueness conditions, and a further assertion, in the second conjunct, which is made about its subject independently of whether the first conjunct is true or false, and so about something that exists eternally. For the same point is central to understanding how such eternally real objects are accessed — which is a seemingly perennial difficulty with Platonic entities. Paradigmatically the situation is represented again in the epsilon variant to Russell's analysis of 'The king of France is bald'. For the first conjunct in the second expression above is itself equivalent to a conjunction:

$K\epsilon x(Kx \ \& \ (y)(Ky \supset y=x)) \ \& \ (y)(Ky \supset y=\epsilon x(Kx \ \& \ (z)(Kz \supset z=x)))$.

So access to the individual $\epsilon x(Kx \ \& \ (z)(Kz \supset z=x))$, i.e. the king of France, is provided entirely by means of the linguistic act of supposing there is a sole king of France, and through its then being invariably possible to cross-refer to the same individual from within further assertions. Eternal objects, in this way, are simply subjects of discourse.

In particular one must remember, at this point, that ordinary proper names are disguised descriptions, and so one must distinguish the individual from the associated property. Thus we may say that $p = \epsilon x(x$ is called 'Pegasus'), just as $r = \epsilon x(x$ is called 'Russell'). And then both p and r necessarily exist, even though $\neg(\exists x)(x$ is called 'Pegasus'), while $(\exists x)(x$ is called 'Russell').

5

There are a great many aspects of fictions that the foregoing throws considerable light on. Thus the consistent propositional base for truth, for a start, shows that no departure from classical two-valued logic is necessary to understand fictions. Some statements, namely those relating to fictions, certainly have no determinate, or 'factual' underpinning. But using the sort of choice

formalised in the epsilon calculus we can settle on a truth value, even arbitrarily, and that is enough to show that no many-valued logic, or supervaluation,
need come in. The shift from a sentential to a propositional account of indirect
speech, as well, relates to another very basic question about literary fictions:
about whether one takes into account translations of literary works. If Conan
Doyle said that Sherlock Holmes lived in Baker Street, and one wants to win
a bet on the question of what he said, does one have to speak English and say
exactly 'Sherlock Holmes lived in Baker Street' rather than some translation
of this? On a strict 'say-so' semantics this might be the case, but what Conan
Doyle said, in the appropriate sense, is not a matter of direct quotation, since
reported speech is involved. A more substantial question concerns whether one
should just say that Sherlock Holmes lived in London, or is it more proper to
mention that Conan Doyle, or some English writer said this? We now have a
clear answer, since we now have unearthed the formal expression for statements
of the latter kind, namely the expression '$V[p, i] = 1$' read as saying that it
is true in world i that p. For if one says merely 'p', by contrast, then one
is just playing along with the storytelling; joining in the fiction, or pretence.
A different speech act is then involved other than assertion, even though an
indicative sentence is uttered. But a more objective report is generally still
available, in the above operatorial form, with 'i' being an index for the author,
or collection of authors, in question. And that also settles a question about the
difference between Sherlock Holmes and the present king of France. For no one
has written a story about the latter (or, at least not a well known one), which
means that no report is available, which would give the authority for the story.
So one is left, in the latter case, to make up some fiction spontaneously, and
say something, non-assertively, of the form '$F\epsilon x(y)(Ky \equiv y=x)$'.

The intrusion of further elements from real life into known stories, as in the
case 'Sherlock Holmes had tea with Gladstone', for instance, can be handled in
the same way, since even if Conan Doyle did not say it, we could still make up
such a story ourselves. But the interplay between real life and fiction is more
complicated than this, with some statements involving fictions being proper
parts of real people's histories. This would be the case with 'Kinglsey Amis
admires James Bond', for instance, to which we might also add the hoary old
philosophical chestnut 'Ponce de Leon was looking for The Fountain of Youth'.
But the bare form,

X was looking for Y,

is invariably about someone's relations to some thing in this world, whether or
not there is, or is believed to be, anything of the kind looked for, as we shall
see with respect to the round square in a moment. For, again, all individuals
necessarily exist and so exist in all possible worlds, while what might distinguish
this world from another is entirely another matter: whether such individuals
'live up to their name', i.e. satisfy all that is said about them. If Ponce de
Leon were to say 'There is no such thing as The Fountain of Youth, but I am
still looking for it' we would maybe judge him crazy. But formally it is just a
matter that there is no point in making any effort to find ϵxFx if one knows
that $\neg(\exists x)Fx$, since an arbitrary nomination of which thing 'ϵxFx' is to refer
to is enough, then, to locate it. In the reverse case, certainly, we have a more

normal, rational situation, but then a fuller presentation of that sort of case than just 'Ponce de Leon was looking for the Fountain of Youth', would be along the lines 'Ponce de Lion knew (or believed) there was such a thing as a sole Fountain of Youth, and he was looking for it'. And whether there is, or someone believes there is such a thing as a sole Fountain of Youth clearly brings in other things than the pure relation of looking for.

Now Graham Priest has recently given an account of 'intentional' predicates and operators which it is useful to compare the above with at this point. There are a number of things common between our two accounts, which I shall first list; but mostly there are very wide divergences. Priest uses the epsilon calculus as I do. He says (Priest 2011 243):

> If something satisfies the condition $A(x)$ in the actual world then $\epsilon x A(x)$ refers to one such (contextually determined) thing. Hence $A(\epsilon x A(x))$ is true. If nothing satisfies this condition then $[\epsilon x A(x)]$ refers to some other, contextually determined object.

As a result he has a fair understanding of the required account of the round square, $\epsilon x (Rx \,\&\, Sx)$. He says (Priest 2011, 243):

> Assuming the actual world is consistent, then nothing satisfies the condition 'x is round and square'. The description, then, will refer to some contextually determined object that does not satisfy this condition. What properties this object has depends, of course, on what object actually]emphis picked out in the context. The fact that there is no determinate answer to the question of what properties it has is hardly objectionable, therefore any more than any other case of contextual determinacy.

Nevertheless he gets into difficulty with, amongst other things, what he calls 'non-existents.' He says (Priest 2011, 238, 243):

> Every world contains the same domain of objects, D. At each world an object may or may not exist. Thus there is a monadic existence predicate whose extension at a world is the set of thing[s] that exist there

> So suppose that a term denotes a nonexistent object. What more can be said about its properties? For a start, it cannot have existence entailing properties, by definition. So it cannot be on top of the Berkeley Clock Tower, and I cannot kick it.

But the round square, in the above sense, clearly *can* be on top of Berkeley Clock Tower, and one *might* kick it, if $\epsilon x (Rx \,\&\, Sx)$ was chosen carefully enough. So Priest's account clearly needs to be amended, and in fact the necessary revision of it is very substantial indeed.

What, in summary, is right and what is wrong with Priest's account in comparison with the above? Well, his first error is that he defines propositions in terms of possible worlds rather than in terms of translations. This derives from the over-riding expectation in his culture that there be a mathematical analysis of logic, rather than a linguistic one. But then he includes 'impossible worlds' as well as possible ones, and propositions that may be both true and false, in tune with his well-known views about paraconsistency. This derives from his attachment to Tarski's theory of truth, and his lack of attention to the crucial differences between this and Horwich's theory. But that attachment is also connected with his endorsement of standard, Kripkean modal logic, along with the usual possible-world formulations of its general intensional extensions, through his failure to read 'V[p, i] = 1' as part of the object language, i.e. as just another operator, and his lack of questioning about the yet-to-be-explained 'accessibility relations'. As well, Priest's account includes a view of individuals such that they may or may not exist; and he has contingent identities in addition to necessary ones. The point is not that Ockham's Razor is sufficient to eliminate all these unneeded beasts; it is the much more substantive one that these monstrosities are based on confusions, and are therefore incorrect. They are fictions in a surreal fantasy story we have been reading, and we must now return to the actual world, and regain the use of our senses. In particular we have seen that on the proper view of truth, there are no true contradictions, that modalities may iterate without any transworld 'seeing', and that hyperintensional relations are not relations with possible worlds. In addition we have seen that the same individuals exist in all worlds, making, at least, Priest's common domain D correct, but also making all identities between individuals necessary. Contingent 'identities' can only arise using something like Russell's iota term symbolism, in which an individuating description is masquerading as an individual term. With that understood then what has contingent existence are some properties, not any individual in which those properties are instantiated.

Priest shows he has some grasp of the required difference between individuals and their individuating descriptions. But he is held up through his belief that existence is to be formulated as a predicate of individuals, making some but not all individuals 'exist' at a world. So he misses the fact that contingent existence is a matter of the instantiation of properties, and the related fact that somethings looking like identities between individual terms are instead talking about the properties of an individual. He says (Priest 2011, 239):

> *Prima facie*, it would appear that Oedipus desired Jocasta, but did not desire his mother, even though Jocasta is his mother. This is not a tough problem, however. Oedipus *did* desire his mother. He just did not realize that Jocasta was his mother. Of course he realized that Jocasta was Jocasta. Hence substitutivity of identicals, we may suppose, holds within the scope of intentional predicates.

He therefore thinks that substitutivity of identicals, while it holds for intensional predicates like 'desire' breaks down with intensional operators like 'realise that'. What Oedipus did not realize, however, was not an identity between individuals, but the property of an individual. What he failed to realise was that $j=\iota x Mxo$, not that $j=\epsilon x Mxo$, where 'Mxy' is 'x is a mother of y' (c.f. Slater 1992a). He realised that $j=j$, and so realized that $j=\epsilon x Mxo$, because $j=\epsilon x Mxo$, and substitutivity of identicals holds even in operator constructions. But he could realize that j=εxMxo without realizing that j=ιxMx, i.e. that (x)(Mx ≡ x=j) through thinking, for instance, that he had a mother who was not Jocasta.

More particularly, with his view of existence, Priest misses just how true it can be, in real life, that we pity Anna Karenina herself (c.f. Slater 1987). For while he wants to say that intensional relations can exist with things that do not exist, and allows that we can say 'I pity Anna Karenina', he does so while realising that 'Sherlock Holmes', and so presumably 'Anna Karenina', is merely a disguised description (in Russell's terminology). So the relation of pity we have is a more general one to something with appropriate characteristics, showing that pitying Anna naturally involves pitying other things with similar descriptions. But that allows we can pity even *her*, if the associated epsilon term, encapsulating the full description, is chosen appropriately. In the actual world the referent of this epsilon term 'εxAx' is strictly arbitrary since no-one has all the required characteristics, i.e. ¬(∃x)Ax. But still we might locate that referent as a specific person amongst Anna's counterparts; for instance an actor on the stage who we see as her.

These points also relate centrally to the debate over emotional responses to fictions. For one must remember that many readers, film viewers, etc. believe the fictions they are involved with are actual, or are similar enough to actual events, or are quite likely true etc., and it is on that kind of basis that understandable human relations might be developed with what would otherwise be just arbitrary entities. Formally some such further presumption must be added before we can have an emotional relation with fictions, otherwise the much-discussed paradox of fiction would arise (Radford 1976). For it is not the case that such full emotional relations are experienced without belief (even with non-fictions). Fictions, on their own, are 'dead', and so the possibility always is there of viewing them entirely 'aesthetically', i.e. without any belief in their (Aristotelian) existence.

6

Why then is Logic not Mathematical? There are a number of heterogeneous specific facts I have demonstrated that show that Logic is a literary pursuit instead. First is the fact that reflexive predicates are not to be handled like binary functions, which has very large repercussions to do with the difference between identity and equivalence, as well as resolving many hard puzzles in Set Theory and predicate logic (also the Lambda Calculus, as we shall see). Second is the fact that propositional identity is a matter of synonymy, and translation, not logical equivalence. Whether or not there is any hope of going beyond bare 'translation', and defining some kind of 'structural isomorphism' between different expressions for the same proposition in different languages is an hypothesis some have entertained, but that is an empirical matter which must measure up to facts about the variant structures of synonymous sentences in those different languages. Third is the irrelevance of mathematical 'accessibility relations' in the understanding of properties of propositions, such as being necessary, being contingent, being believed by a, and being known by b to be believed by a etc. Fourth is the inevitable indexicality, and therefore pragmatic use involved in the expression of propositions, and thereby the understanding of Truth. And these contextual features also need to be understood, fifthly, to grasp what individuals are, because of the necessary choice involved in using referring terms. But that leads, sixthly, to the fact that individuals are features of discourses that suppose there are things with the associated character, and so which, independently of logic, may or may not be fictional. The Logical Empiricists, from whom so much of modern logic has descended, thought only of physical objects as contingent beings, but the coherence of the propositions made by some sentences can, indeed must be assessed quite independently of whether they are about actually existing things. Formally that leads to a better understanding of the notion of Existence, but the consequent move to a detailed consideration of the place of fictions in human lives much more fully shows how the focus has shifted towards Literature.

APPENDIX I

THE CENTRAL ERROR IN THE TRACTATUS

I.A

Robert Fogelin claimed there was an error in the logic of the Tractatus. I first cover his point here before going on to show that any error in this area derived from an even more fundamental one. Correcting that further error, moreover, does more than correct the logic of the Tractatus: it has repercussions for the metaphysics and theory of value found there, in line with later developments in Wittgenstein's philosophy. In what follows I use the Tractarian numbers to indicate the paragraphs spoken about.

I.B

The logic of the Tractatus was very influential, but there is a difficulty with it which was pointed out by Fogelin. In the Tractarian predicate logic, and specifically in its account of generality, Wittgenstein continues to employ his operator 'N', the form of Sheffer's Stroke he had used in connection with his propositional logic. He says, amongst other things (5.52) 'if ξ has as its values all the values of a function Fx for all values of x then $N(\xi) = \neg(\exists x)Fx$'. Fogelin focusses on a problem concerning multiple generality, which arises because 'N' tries to replace a quantifier without using an index to show which variable is being quantified over. Others (Geach and Soames, for instance) have tried to modify 'N' appropriately, to allow for this, but Fogelin has stuck to the text (c.f. Cheung 2000). Fogelin points out that, in order to generate, say, '$\neg(\exists x)\neg(\exists y)Fxy$' it might seem one could say that if ξ has as its values all the values of the function '$\neg(\exists y)Fxy$' for all values of x then $N(\xi) = \neg(\exists x)\neg(\exists y)Fxy$. But since '$\neg(\exists y)Fxy$' would itself have to be $N(\zeta)$ where ζ has as its values all the values of a function of two variables Fxy, this is either disallowed by the mention of functions of merely one variable in 5.52, or, as Fogelin shows (Fogelin 1976, 71) generates instead '$\neg(\exists x)(\exists y)Fxy$'. So one would have to consider an infinite enumeration

$N(\neg(\exists y)Fay, \neg(\exists y)Fby, \neg(\exists y)Fcy...)$.

But while enumerations are allowed in $N(...)$ (5.501) Fogelin points out (Fogelin 1976, 72-3) that only a finite number of truth-operations were condoned (5.32). Other remarks in the Tractatus suggest that Wittgenstein was prepared to allow infinite totalities (see, e.g. Ramsey 1978, 158; Marion 1998, 34), but paragraph 5.32 is final against an infinite number of truth-operations. One might want to say that the operation 'N' is only applied once even in an infinite case: the application of 'N' to (p, q) produces '$\neg p \& \neg q$', in which there are as many negations as there are arguments, but the whole transformation is just one application of 'N'. But negation is itself a truth-operation, and there is a further point: the kinds of enumeration envisaged in paragraph 5.501 were finite lists, and opposed there to the giving of a function Fx whose values are the propositions in question. It was Wittgenstein's view that there is no way of writing down an infinite list other than through such a variable form, or formal law. Ramsey disagreed, but Wittgenstein flatly rejected 'unordered generalities' (see, for instance, Fogelin 1983). Thus in the above case, we can see the way the enumeration is intended to develop by replacing 'a', 'b', 'c' with the variable 'x', even though that specific variable form of expression was not available, in its standard use, within the Tractatus, as Fogelin showed.

The impossibility of giving such a function in the above case would have been removed if the Tractatus language had included Skolem functions, for instance. With them in the language we might be able to write '¬(∃y)Fxy' as '¬Fxg(x)' for a certain function 'g(x)', and then apply paragraph 5.52 directly to a function of one variable, to get '¬(∃x)¬Fxg(x)', and thereby '¬(∃x)¬(∃y)Fxy', as required. Whatever the virtues of Geach's or Soames' emendations, however, there are quite fundamental reasons why the explicit language of the Tractatus could not accommodate Skolem functions, as we shall now start to see.

For there is another error in the logic of the Tractatus, which Fogelin missed, even though his discussion of the logic of the Tractatus was much fuller, and more critical than most. That further error concerns the logic of identity, which, as is well known, Wittgenstein handled or tried to handle using identity of sign (5.53). That requires using variables that are 'exclusive' rather than the standard 'inclusive' ones. Fogelin gives some examples (Fogelin 1976, 66-7). Thus 'There are at least two F's', written with inclusive variables

(∃x)(∃y)(Fx & Fy & x≠y),

becomes

(∃x)(∃y)(Fx & Fy)

with exclusive variables, and 'Somebody likes somebody' is

(∃x)(∃y)(Px & Py & Lxy) v (∃x)(Px & Lxx)

rather than

(∃x)(∃y)(Px & Py & Lxy).

Fogelin goes on (Fogelin 1976, 67):

> It seems obvious - though a proof for this is needed that Wittgenstein's method can shadow Russell's making numerical assignments to things already described under some other non-logical predicate. But Wittgenstein's procedures will not produce counterparts for what we might call pure occurrences of the identity sign, i.e. occurrences of the identity sign governing individuals not previously qualified by some non-logical predicate. It thus seems that the only occurrences of an identity sign that are not eliminable by Wittgenstein's procedure arise in expressions that Wittgenstein wishes to exclude from the language.

Fogelin, at the end here, had in mind paragraph 5.534, which says propositions like 'a=a' and '(∃x)(x=a)' cannot even be written down in a correct conceptual notation, and paragraph 4.1272, which says one cannot say 'There are objects',

'There are 100 objects', etc. Thus the example Fogelin considers (Fogelin 1976, 66):

$(\exists x)(y)(y=x)$,

we might read as 'there is just one object', but this is a pure identity proposition, without any non-logical predicate, and so it will not be expressible, Fogelin thinks, in Wittgenstein's preferred logical language. But Fogelin seems to have been unaware of work by Hintikka in 1956, in which the needed proof he spoke about above was in fact given. Hintikka produced a general translation of all propositions available in standard predicate logic with identity (which there were expressed using inclusive variables, of course), so that only exclusive variables were involved. He wrote the exclusive translation of a formula 'L' as '$\phi(L)$', and summed up (Hintikka 1956, 235):

> We have already observed that if L is closed, $\phi(L)$ does not contain any identities; and it is well known that in the predicate calculus we have no need to consider formulae which are not closed. Hence the result we have just proved shows that everything expressible in terms of the inclusive quantifiers and identity may also be expressed by means of the...exclusive quantifiers without using a special symbol for identity.

In a footnote he immediately adds: 'Wittgenstein was, hence, right in saying that the identity sign is not an essential constituent of logical notation'. Unfortunately for Fogelin, and in a different way for Hintikka, Hintikka's mode of re-expression does too much: for it also gives translations of Fogelin's pure identity propositions. This is implicit in what Hintikka says above, since pure identity propositions are just some of the propositions in standard predicate logic with identity. So it is surprising that Hintikka did not consider, or at least comment on, what Wittgenstein has to say in paragraphs 4.1272, and 5.534. But for Fogelin these remarks have more significance, and yet they cannot be supported in the way he supposes. To make the matter entirely clear I will give Hintikka's translation in Fogelin's chosen case above. Hintikka's translation process, for a proposition K with two free (inclusive) variables x and y, follows the general plan of the translation for 'Somebody likes somebody' (Hintikka 1956, 231-2). He first finds two propositions K0 and K1 which contain no identity propositions, such that K is equivalent to

$(K0 \ \& \ (x{\neq}y)) \ v \ (K1 \ \& \ (x{=}y))$.

Then the translation of $(\exists y)K$ becomes

$(\exists y)K0 \ v \ K1(x/y)$,

using exclusive variables. Fogelin's case is '$(\exists x)\neg(\exists y)\neg(x{=}y)$', and merely requires introducing arbitrary tautological (T), or contradictory (F) propositions for the appropriate K0 and K1. Thus '$\neg(x{=}y)$' is equivalent to '$(T \ \& \ \neg(x{=}y))$ v $(F \ \& \ (x{=}y))$', and so '$(\exists y)\neg(x{=}y)$' becomes '$(\exists y)T \ v \ F(x/y)$'. Its negation '$\neg(\exists y)\neg(x{=}y)$' becomes '$(y)F \ \& \ T(x/y)$', and finally 'there is only one object' becomes

$(\exists x)((y)F \ \& \ T(x/y))$.

The latter is not the formal contradiction it may seem, note, since it returns to the inclusive form via '$(\exists x)(y)(\neg(x{=}y) \supset F)$' (see Hintikka 1956, 230). So, apparently, formal concepts are expressible: we can say 'There is just one object' in a language without the identity symbol.

I.C

What are we to make of this? Was Fogelin too hasty in simply endorsing Wittgenstein on 'There are 100 objects' etc? Was Wittgenstein completely wrong about formal concepts - which would be a really major error? Certainly Hintikka seems to be misleading if he intended in his footnote to defend the whole of the Tractatus in this area. But what defence of Fogelin's separation of pure identity propositions could there be, given Hintikka's quite general translation?

One aspect of this matter is that there are two sorts of claims possible regarding pure identity propositions: 'they cannot be said in a correct conceptual notation' could mean 'there is no exclusive representation of them', and it could mean 'they are nonsense propositions, since they have the non-factual status of tautologies and contradictions'. Hintikka showed that 'there is just one individual' can be re-expressed in the exclusive notation; but Fogelin was certainly more concerned elsewhere with the status of propositions, as, for instance, in his discussion of fully general propositions (Fogelin 1976, 60-4).

Ramsey illustrates the two major aspects of this issue in the following passage (Ramsey 1978, 211):

> Next let us take 'There are at least two individuals' or '$(\exists x,y)(x \neq y)$'. This is the logical sum of the propositions $x \neq y$ which are tautologies if x and y have different values, contradictions if they have the same value. Hence it is the logical sum of a set of tautologies and contradictions; and therefore a tautology if any one of the set is a tautology, but otherwise a contradiction. That is, it is a tautology if x and y can take different values (i.e. if there are two individuals) but otherwise a contradiction.

Here Ramsey, because of his rather different concept of identity, is happy with 'there are two individuals' not only being expressed, but also being a condition which might be true or false, even though it itself is going to turn out to be either a tautology or a contradiction, and so not conditional at all. Wittgenstein was, on this basis, not entirely wrong about formal concepts: they are expressible, but their expression must result in vacuous tautologies or nonsensical contradictions.

We have not, however, seen the end to the significance of Hintikka's result, nor to the full assessment of paragraphs 4.1272, and 5.534. For we have not attended to the initial caveat in Hintikka's statement of his result: he said it applied only to closed formulas in the predicate calculus with identity. Hence ordinary identity statements of the form 'a=b' are not covered, and must be given a separate treatment. For one thing, they should not be lumped with other pure identity propositions as in paragraph 5.534. More significantly, despite Ramsey wanting to say they were tautologies or contradictions above, they are the only statements which come into Fogelin's special category: it is just plain identity statements which cannot be expressed in Wittgenstein's approved way, using exclusive individual terms, in a language lacking identity. The fact that they are not going to be expressed is just the content of paragraph 5.53, in which Wittgenstein announces his intention to use an exclusive language and abandon the identity symbol. The singularity of this decision, we shall now see, underpins a much broader set of opinions in the Tractatus, to do not just with the expression of generality flagged before, but also the independence of atomic propositions, the place of the synthetic *a priori*, and even the place of the will, and questions of value.

I.D

When Wittgenstein asserted 'Identity of object I express by identity of sign and not by using a sign of identity. Difference of objects I express by difference of sign.' he was putting himself in the place of God. We can only use the resulting exclusive language without identity when we know the identity relations between objects expressed inclusively. It is a privileged language in which the ignorance cancelled by 'x=y' or 'x≠y', in their inclusive sense, cannot arise. The inclusive forms are used when we are not in a fully knowing state, and so when there is a question still to be asked. Wittgenstein had read Frege, and knew that if x=y then to say that x=y is to say that x=x, while if x≠y then to say that x=y is to say that x≠x. But the consequences 'x=x' and 'x≠x' are not evident in the language expression 'x=y', which is what we mere mortals have to deal with, in ignorance of the facts. The problem with 'x=y', for Wittgenstein, was that it was certainly a necessary truth or falsity, and yet its truth or falsity was not evident in its face, and so it was not what was then called 'analytic *a priori*'. By removing identity statements from his logical system, Wittgenstein in the Tractatus was therefore on the way to eliminating one major portion of what was then understood as the 'synthetic *a priori*' from logic (c.f. Ayer 1958, ChIV). As Kripke later showed, however, (Kripke 1972), what others might now prefer to call the '*a posteriori* necessity' of identity statements is still there, and should be acknowledged.

One has to be careful, when identifying identity propositions. Ramsey understood this very well (Ramsey 1978, 167-8):

> ... these are not real propositions at all; in 'a=b' either 'a', 'b' are names of the same thing, in which case the proposition says nothing, or of different things, in which case it is absurd. In neither case is it an assertion of fact; it only appears to be a real assertion by confusion with the case when 'a' or 'b' is not a name but a description. When 'a', 'b' are both names, the only significance which can be placed on 'a=b' is that we use 'a', 'b' as names of the same thing or, more generally, as equivalent symbols.

Ramsey here was clearly alluding to the analysis of descriptions provided by Russell: Russell had allowed 'identities' to be formed like 'a = $\iota x D x$', but this is not a proper identity, being re-expressible '(y)(Dy ≡ y=a)'. After such cases are sieved out we are left with noms de plume, and aliases, for instance. But we can quite easily be ignorant of someone's alias, hence the usefulness of (inclusive) identity propositions, and the extreme oddity of the position Wittgenstein was putting himself in, by trying to avoid them. Is such an identity statement an 'assertion of a fact'? If we are ignorant of anything it is merely the use of the names, since 'a=b' is empty of factual content about the world, as Ramsey says. This is also shown by its origin in pure acts of will, providing verbal alternatives for the same thing. But with the provision of this choice we immediately get the non-independence of certain elementary propositions, since we can express Leibniz' Law. Given a=b we can say

Pa ≡ Pb,

for any 'P', and even if 'Pa' then merely says the same as 'Pb' another way, still we have two elementary propositions which are equivalent. Upon Wittgenstein's return to philosophy in 1929, of course, he came to question his earlier dictum about the independence of elementary propositions. But instead of the much larger body of examples he considered, which convinced him elementary propositions need not be independent, he might just have considered how Leibniz' Law originates, through having a choice, as above.

Pairs of elementary propositions between which there is simply an entailment arise when other identities are settled upon, associated with the Skolem functions mentioned before, in connection with generality. We considered then how '¬(∃y)Fxy' might be replaced with '¬Fxg(x)', with 'g(x)' a Skolem function. In the simpler case of one variable, '(∃x)Fx' is equivalent to 'Fs' for a Skolem constant s, and so, since

Fa ⊃ (∃x)Fx,

we can obtain

Fa ⊃ Fs,

making these two elementary propositions also non-independent.

I.E

A formulation of Skolem functions and constants is provided by Hilbert's epsilon calculus. And that leads us to see how correcting the above error in the Tractatus is connected with later developments in Wittgenstein's philosophy. For '(∃x)Fx' is logically equivalent to 'FεxFx', where the referent of the epsilon term 'εxFx' is chosen from amongst the F's, if there are any, and from the universe at large if not (Leisenring 1969, 19). Then not only might '¬(∃y)Fxy' be replaced with '¬Fxg(x)', more particularly it may be replaced with '¬FxεyFxy'. In the constant case, if s = εxFx, then we get the implicational relation given above. An epsilon term, note, is a logically proper name, i.e. a 'complete' term for an individual, unlike the iota terms considered before, and so 'εxFx = s' is not further analysable, and comes into Ramsey's non-descriptive class.

The epsilon term 'εxFx' Hilbert read 'the first F', because it refers to a paradigm object with the property in question, such items being guaranteed on account of the relation between the epsilon axiom

(∃x)Fx ⊃ FεxFx,

and the predicate calculus thesis

(∃y)((∃x)Fx ⊃ Fy).

'If anyone is just then Aristedes is just' was the original example used by Hilbert when he introduced epsilon terms (Reid 1970, 270, see also Copi 1973, 110). Likewise 'If anything is a material object then Moore's hands are' would be a good illustration of what is involved in applying the epsilon axiom with 'Fx' meaning 'x is a material object', and '∃xFx' being chosen to refer to Moore's hands. Such paradigms as the standard metre, and colour samples, were notable features of Wittgenstein's later understanding of concepts, as Fogelin has explained (Fogelin 1976, 112f): 'If anything is a metre long, then the standard metre is a metre long' insists on the centrality of a certain extra-linguistic object (related to Wittgenstein's notion of a 'simple' by Fogelin) in connection with the measurement of length.

These kinds of cases, in fact, inspired 'The Paradigm Case Argument', which was widely applied in the period of Linguistic Philosophy as a general means of understanding all concepts. One could not doubt that a paradigm had the respective property without bringing into question the whole concept involved, so the temptation was to settle any question simply by affirming the paradigm. In its heyday The Paradigm Case Argument was even thought to answer not only most of the traditional problems in epistemology, but even those in axiology, to do with our power to make things happen (Black 1958), our freedom (Flew, 1966, Ch1), and our respect for moral value (Bambrough 1979, Ch2). Indeed Bambrough drew an explicit parallel between Moore's case in epistemology and his case in axiology, and, as above, ethical, social and political values are involved in 'If anyone is just then Aristedes is just', the original example used by Hilbert.

The generations of philosophers subsequent to the Linguistic Philosophy period have not understood things this way, due in part to observing an important flaw in The Paradigm Case Argument (c.f. Passmore 1961, 114f). For Aristedes still might be unjust — the only consequence being that that would make his nickname 'The Just' an ironic misnomer, and be the end of justice:

$$\neg F\epsilon xFx \supset \neg(\exists x)Fx.$$

Likewise, while Moore's hands might be accepted as paradigm material objects, that properly only leaves us with the conditional truth above: Moore's hands are material objects if anything is. So the traditional epistemological problem was not solved, as Linguistic Philosophy had assumed, and more philosophy seemed to be needed, to defeat Skepticism in the area. But that move to more argument itself involves a misconception of the situation, from a Wittgensteinian perspective. Specifically it misrepresents the role of the will in the matter. For there is no proof of the existence of the external world which could improve on Moore's procedures. There is a demonstration of it, in the way that Moore proceeded, but responding to that demonstration is a matter of human choice, a matter of 'how we act' (Shiner 1977/8). Likewise in Bambrough's case of giving an anaesthetic to a child who needs an operation: there is no proof that this is right, but appeal to this sort of case is reason's last resort, since if someone does not respond normally to children in pain they have lost touch with the concept of morality. If a person falls in line with Moore's and Bambrough's clear illustrations, then they start to acquire the standard concept of a material object, and the standards of morality. But humans are, logically, quite free not to fall in line, even if then, as the reversal of the conditional shows, nothing will be material, or moral for them, in the usual sense. They then 'act differently', and simply live a form of life at odds with the rest of us. It is a personal, existential choice, in other words, and there is no final argument against it.

Correcting the central error in the Tractatus therefore does more than correct the logic of the Tractatus: it has repercussions for the metaphysics and theory of value found there, in line with later developments in Wittgenstein's philosophy.

APPENDIX II

FREGE'S HIDDEN ASSUMPTION

II.A

This chapter is concerned with locating the specific assumption that led Frege into Russell's Paradox. We shall see that his understanding of reflexive pronouns was weak, for one thing, but also, by assimilating concepts to functions he was misled into thinking that one could invariably replace a two-place relation with a one-place property.

It is well known that it was Russell's Paradox that alerted Frege to the trouble with his system, for substitution of the set-abstract for 'x' in

$x \in \{y : y \notin y\} \equiv x \notin x,$

produces a contradiction. It is less well known that there is not the same trouble with

$<x,x> \in \{<y,z> : y \notin z\} \equiv x \notin x.$

For substitution of the set-abstract for 'x' here does not produce a similar contradiction (c.f. Slater 1984). Substituting the set abstract for 'x' in the first case yields something of the form

'$a \in a \equiv a \notin a$',

but the comparable substitution in the second case merely produces something of the form

'$<b,b> \in b \equiv b \notin b$'.

What this suggests is that '$x \notin x$' cannot be analysed as involving simply a predicate of x rather than a relation between x and x, the larger moral being that not all relations between a thing and itself can be a matter of that thing falling under a concept, i.e.

$\neg(R)(\exists P)(x)(Rxx \equiv Px).$

This is defended further in what follows, but it is what might have led Frege to think otherwise which is the main interest in the present chapter.

II.B

In fact Carnap's notes on Frege's lecture course on the *Begriffsschrift* show precisely where Frege went astray. Other passages in Frege might also be quoted, of course, but these lecture notes, which have recently been published in English, are particularly clear in this regard. For there is an argument in them about concepts and relations, and specifically about the possibility of generating certain (one-place) concepts from certain two-place relations. One can put the problem highlighted by Russell's Paradox directly in terms of relations and concepts, to bring it closer to these kinds of expression. Remembering the general format of lambda abstraction reduction is $\lambda yFy(a) = Fa$, the trouble with the Paradox of Predication,

$(\exists P)(x=\lambda tPt\ \&\ \neg\lambda tPt(x)) \equiv \lambda yQy(x),$

is that one seems to be required to equate the 'λyQy' on the right with

$\ \ \ '\lambda y(\exists P)(y=\lambda tPt\ \&\ \neg\lambda tPt(y))',$

i.e. with the expression formed by abstracting the 'x' from the left hand side of the equivalence, to produce a concept of x. But this does not work. For there is then a contradiction when 'λyQy' replaces 'x'. One gets, because

$(\exists P)(\lambda yQy=\lambda yPy)$

is guaranteed, something of the form

$\ \ \ \neg c(c) \equiv c(c).$

On the other hand, there is no similar paradox if one abstracts from each 'x' separately, i.e. with

$(\exists P)(x=\lambda tPy\ \&\ \neg\lambda tPy(x)) \equiv \lambda y\lambda z(\exists P)(z=\lambda tPt\ \&\ \neg\lambda tPt(y))(x)(x).$

In this case, substituting

$\ \ \ '\lambda y\lambda z(\exists P)(z=\lambda tPt\ \&\ \neg\lambda tPt(y))'$

for 'x' does not produce a contradiction. Hence the left hand side,

$\ \ \ '(\exists P)(x=\lambda tPt\ \&\ \neg\lambda tPt(x))',$

cannot be analysed as involving simply a concept of x rather than a relation between x and x. A paradox only arises when taking the two argument relational expression on the left (with the two arguments identified the same) to be equivalent to a single-subject with constant predicate expression, as when there was just

$\ \ \ '\lambda y(\exists P)(y=\lambda tPt\ \&\ \neg\lambda tPt(y))(x)'$

on the right.

With respect to the Paradox of Predication, we therefore see that, while

$\ \ \ $'x is a property which y does not possess'

expresses a relation between x and y, that does not mean that, if 'y' is replaced by 'x', the result is a one-place property of x. And it is also very clear why this is so. For the predicate in the diagonal case,

$\ \ \ $'is a property which x does not possess'

is itself something that varies with the subject specified. What is predicated of a would be that it is a property which a does not possess, but what is predicated of b would be that it is a property which b does not possess. In other words, the general predicate can be taken to be 'is a property which it does not possess', and this contains a pronoun, which is a contextual item with no direct representation in a context-free language. This pronominal predicate is functional, in other words, and the nearest one can get to any concept it expresses, in a language without pronouns, is:

$\lambda y\lambda z(\exists P)(z=\lambda tPt\ \&\ \neg\lambda tPt(y))(s),$

with 's' not entirely free, but limited to repeating the subject of the sentence. Alternatively, by taking the second entry of 'x', in the diagonal case, as the subject one could take the general predicate to be 'does not possess the property it is', and the nearest one could get to any concept this expresses, in a context independent language would be

$\lambda z\lambda y(\exists P)(z=\lambda tPt\ \&\ \neg\lambda tPt(y))(s).$

The attempt to construe these predicates as expressing a functional concept of their subject becomes needless, however, once the required subject term is actually attached, since the whole sentence is then revealed to be simply, though irreducibly relational, by each time being the analysis that was paradox free, namely:

$\lambda y \lambda z (\exists P)(z=\lambda tPt \ \& \ \neg \lambda tPt(y))(x)(x).$

II.C

What Frege first missed with reflexive forms was the functionality of such pronouns. In the *Begriffsschrift* he says (Frege 1972, 127):

> The proposition that Cato killed Cato (can be considered in three ways, involving three different functions). Here, if we think of 'Cato' as replaceable at the first occurrence, then 'killing Cato' is the function. If we think of 'Cato' as replaceable at the second occurrence, then 'being killed by Cato' is the function. Finally, if we think of 'Cato' as replaceable at both occurrences, then 'killing oneself' is the function.

But 'killing s', with 's' a pronoun, is not expressible in a context-free language, and, in addition, there are, in this case, other functions expressible in such a language which Frege does not mention: the two-place ones where the same name may (but need not) be put in both places: 'killing' and 'being killed by'. So why did Frege think that by abstracting 'Cato' from both occurrences in 'Cato killed Cato' one obtains a one-place function rather than a two-place one?

Here is the passage in Carnap's notes that provides the answer. Frege says (Frege 2004, 155):

> A function of two arguments, e.g. x-y, can be transformed into a function of one argument in two different ways, either by saturation (x-2), or by identifying the two argument places (x-x). Functions of two arguments that always have a truth-value as value are relations. Therefore we can transform the relation x>y into a concept, e.g. x>0 (the concept of a positive number). Or we can form the concept x>x.'

If Frege could get the reflexive concept at the end, from the relation he started with, then a comparable derivation would produce a concept of x from the relation between x and x, on the left in the Paradox of Predication. So clearly this cannot be done.

How could Frege have missed the fact that only a reflexive relation, and not a reflexive concept is derivable? Clearly it was Frege's background in Mathematics that got him into trouble. Specifically, if there is no derivation of the supposed concepts from the given relations (both in the case of 'x¿x', and in the case of the Paradox of Predication), then Frege must have been working entirely on the basis of his understanding of mathematical functions — which is also nicely illustrated in the passage above. For there is no doubt that, given any function of two variables, $f(x,y)$, one can invariably obtain a function of one variable, by identifying the two arguments: $f(x,x)=g(x)$. One important case where this undoubtedly happens is in Cantor's diagonal procedure, for instance. But the seeming parallel case with relations and predicates, which generates Russell's Paradox, works very differently, as we have seen. So, while it is well known that Frege thought of concepts as functions, the analogy between the mathematical case, and the 'truth-value' case must limp just at this point. The point to note is that the 'truth-value' case involves an equivalence, not an identity, and we now know, from all logic texts subsequent to Frege, that identity is not equivalence. Frege himself had a curious system, which allowed him to conflate, to some extent, identities and equivalences; but this has not been followed, and for good reason. For the expression for an identity, like 'a=b', is between two names, whereas the expression for an equivalence, like 'p ≡ q' is between two sentences. Thus 'if and only if' has quite a different grammar from 'is identical to'. Maybe by 'sentence' Frege meant 'nominalised sentence', since those certainly are referential expressions, and we can, as a result, say 'John's being a bachelor is the same as him being an unmarried male'. But since Frege did say 'sentence' we have every right to correct him. One cannot say, for instance, 'John is a bachelor is identical with John is an unmarried male'.

Sentences are the sort of expression that enters into equivalences, so they are not referring terms which can enter into identities (c.f. Prior 1971, 35), and specifically, therefore, sentences are not referential terms with the same reference as 'The true' or 'The false', as Frege thought. If anything at all like 'Pa=T', or 'Rab=F' holds, it is with '=' as material equivalence, 'T' a tautology, and 'F' a contradiction. And then one has that 'Pa ≡ T' and 'Rab ≡ F' are equivalent to 'Pa' and '¬Rab' respectively, making 'Pa' and 'Rab' quite unlike mathematical functions, and 'T' and 'F' nothing like their values.

On the specific question of a reflexive relation being a function of one argument, certainly one might be able to define a function f(x) such that, say,

Rxx ≡ (f(x) = 1),

¬Rxx ≡ (f(x) = 0).

But not only does that not make the relation the function, also the right hand sides of these equivalences cannot be figured as involving the same predicate of x. For 'f(x) = 0' is not contradictory, but merely contrary to 'f(x) = 1'. If 'f(x)=0' was replaced by 'f(x) ≠ 1' there would be the same predicate of x; but no specific function would then be defined.

The propositional equivalences above though, namely 'Pa ≡ T' and 'Rab ≡ F', maybe still suggest that predicative expressions are functions of some sort. So we must delve deeper. The question in Carnap's case is whether from the relation $\lambda y \lambda x(x > y)$, one can obtain the concept $\lambda x(x > x)$, as well as the concept $\lambda x(x > 0)$. The second reduction is straightforward, since applying the two-term relation to 0 one gets the concept of being greater than 0:

$\lambda y \lambda x(x > y)(0) = \lambda x(x > 0)$.

But the first reduction hits a problem. Proceeding as before one might try

$\lambda y \lambda x(x > y)(x) = \lambda x(x > x)$,

but in '$\lambda x \lambda x(x > y)$' the 'x' is a bound variable, and so the whole is equivalent to '$\lambda y \lambda z(z > y)$', and using that form one merely gets

$\lambda y \lambda z(z > y)(x) = \lambda z(z > x)$,

i.e. the concept of being greater than x. Frege talks about getting his second, reflexive, concept by identifying the two variables, but he cannot be thinking that one can produce his second concept from

'$\lambda y \lambda z(z>y)(x)(x)$',

since while that produces the statement

'$x>x$',

it still does not identify the concept he mentions, because that statement is still analysed as a relation between two arguments, not as involving a single subject with a predicate expressing the concept $\lambda x(x>x)$. Certainly if we could form

$\lambda x(\lambda y \lambda z(z>y)(x)(x))$

we could get the desired

$\lambda x(x>x)$,

but the abstraction of x in that larger form is just as questionable, given that a relation between a thing and itself is not necessarily replaceable by a concept applicable to that thing.

II.D

It might be said that it is anachronistic to use closed lambda terms to try explicate what Frege was saying. The Paradox of Predication, for instance, is not obtainable in Frege's system, since he did not allow the application of a concept to another concept. So he most probably would have resisted the use of closed lambda terms in any explanation of what he meant by transforming the function $x>y$ into the function $x>x$. But if one cannot talk about Frege in a language he would not use, then one cannot criticise him on a scientific basis. One could not tell him, for instance, that sentences are not referring terms, as was done above, since 'for Frege' they are referring terms, and so one's remarks, it might be said, are not about what Frege was talking about, namely 'Frege sentences' which are referring terms, by definition. Popper, however, amongst several others, had a lot to say about this sort of thing in connection with closed societies, and pseudo-science.

Frege, in his article 'Sense and Reference', wanted the 'F' and the 'a', in an elementary sentence such as 'Fa', to be both referring phrases, taking the reference of the whole — a truth value — to be formed from the references of the parts. But only the singular term is referential: both the predicate, and the sentence as a whole, are merely expressive (c.f. Kneale and Kneale 1962, 585-6, Prior 1971, 35, Wright 1983, 21). They are expressive of a concept and a proposition, respectively. A further argument Frege had for his 'truth-value' conclusion rested on what is commonly called his 'slingshot' (see Neale, 1995, 765, and 791-5). But the irony with this argument is that it is plainly invalid if complete individual terms are used for referential phrases (Neale 1995, 795f), and Frege's extensional logic (unlike Russell's, for instance) did employ such complete referential terms. As it stands that point provides merely an *ad hominem* argument against Frege, but I made it very plausible, earlier, that a better representation of referential phrases is obtained using certain other complete terms, namely Hilbert's epsilon terms, and so the inadequacy of Frege's 'slingshot' argument can be argued for much more generally.

The most crucial reason why sentences are not referring phrases, however, arises from the more basic fact mentioned above, that predicates are not refer- ring phrases either. 'For Frege' they were, but Frege's thought at that point, of

course, was what got him into his paradox about the concept *horse*. Following
Cocchiarella (Cocchiarella 1986), we can remove that paradox. For we can distinguish, as before, the concept of being a horse (λxHx) from the predicate 'is a
horse' ('λxHx( )'), and so see that it is the latter, and not the former, which is
unsaturated. That is because 'being a horse' is a nominalised predicate, which
hence is a referring phrase, while 'is a horse' is not nominalised, and contains a
gap that needs to be filled before a complete thought can be expressed. If the
predicate was a referring phrase, and referred to the concept, then that concept would certainly be unsaturated, but also the phrase 'the concept *horse*'
would not refer to it, since that is saturated. Hence there would be Frege's
paradox. But the predicate 'is a horse', while unsaturated, is not a referring
phrase, and what does refer in the area is the nominalisation of a predicate,
such as 'being a horse'. Indeed Frege himself, in his informal language, used
nominalised predicates to refer to concepts, as when talking about 'being killed
by Cato' and the like, in the first quotation above. But his 'official' position
was that non-nominalised predicates had this purpose, so his theory was not
in tune with his practise. Frege lacked a symbol for nominalised predicates in
his formal language, which is what fundamentally led him to the conclusion
that there is an inadequacy in natural language at this point, when it comes to
expressing the semantical facts. For there is no inadequacy in natural language
when it comes to expressing the associated *natural language* semantic facts,
and Cocchiarella has provided a symbolisation separating out predicates from
their nominalisations, so natural language in this area can now be represented
in a properly formal manner. Adopting Cocchiarella's symbolism we thereby
move over to a clearer formal language without Frege's paradox, and with a
clear distinction between predicates and their nominalisations, for a start.

But Cocchiarella's language naturally contains nominalisations of zero-place
predicates, i.e. closed sentences, and misconceptions about such nominalisations also got Frege mixed up about truth. For it is not a sentence 'p', but
its associated 'that'-clause ('λp', see Cocchiarella 1986, 217) which is referential, and the subject of judgements of truth and falsity. Thus it might be
judged true or false that 5 is a prime number, for instance. But while 'that
5 is a prime number' therefore refers, it refers to a proposition, not a truth
value, and so judgements of truth and falsity do not equate something with a
truth-value, but instead predicate a truth-value of a proposition. That is to
say, such judgements are not in the form of referential identities like 'λP5=T'
and 'λP5=F', with 'T' and 'F' 'The true' and 'The false'. They are instead
predicative remarks like 'TλP5' and 'FλP5', with 'T' and 'F' 'is true', and 'is
false', where the lambda expressions obey the propositional schemas: Tλp $\equiv$
p, Fλp $\equiv$ ¬p. Certainly 'is a prime number' is then a function taking as values
propositions which are true or false, but that means it is a *propositional* function, not a truth-value function like those in Frege's 'Function and Concept'
(see, e.g. Frege 1952, 28). As a result, the focus has to be on what sentences
express.

So look at a reflexive case again. If A, B, and C each shave D then they do
the same constant thing — shave D — but if they each shave themselves, or,
in a ring, shave their neighbour on their left, say, then they do not do the same

constant thing, since what they do merely has a common functional expression: shave f(s), where the variable s is the subject. That means that a reflexive predicate is never, in itself, equivalent to a constant one-place predicate — although, contingently, of course, such a pair may be equivalent. They will be, for instance, if the number of objects involved is finite, since they then can be listed, and do not need to be described. Thus if all and only A, B, and C are self-shavers, then 'x is a self-shaver' is materially equivalent to 'x is one of A, B, and C'. But it is not *logically* equivalent to this disjunction, i.e. it does not say the same thing. For the variable in the predicate 'is a self-shaver', namely the pronoun 'self', prevents the whole expressing a fixed property of its subject.

Of course one can put 'A shaved D' differently. In line with the point made right at the start, one can say 'A and D are a shaving (i.e. shaver-shaved) pair' in place of 'A shaved D', and the predicate in that formulation is not functional even in the reflexive case: 'A and A are a shaving pair'. But the latter does not predicate a fixed property of just A, since instead it predicates a fixed property of the ordered pair consisting of A and himself. Focussing on what they express, therefore, we see that a relational expression like 'Rxy' invariably generates thoughts about the two objects x and y, and the reflexive, or diagonal expression 'Rxx', as a result, generates thoughts about x and itself. Certainly the latter thoughts can be taken to be thoughts about the single subject x, as Frege saw with respect to 'Cato killed Cato'. But what Frege missed there was that the two ways in which this can be done each can be expressed with a pronoun, since each of those two thoughts about the subject involve a further function of it. Thus what is thought *about Cato* could be that he killed Cato, i.e. that he killed himself. More generally

$$Rxx \equiv \lambda yRyx(x) \ (\equiv \lambda yRys(x)),$$

and it could be that he was killed by Cato, i.e. that he was killed by himself. More generally

$$Rxx \equiv \lambda yRxy(x) \ (\equiv \lambda yRsy(x))).$$

Frege had no way to differentiate 'killing himself' from 'being killed by himself', but in both cases the pronoun is simply a context dependent replacement for the immediate subject, allowing the two expressions to be differentiated in a partly context-sensitive language as '$\lambda yRys$' and '$\lambda yRsy$'.

The tradition in modern logic has followed Frege in this respect, making a difference from the case with identities and equivalences. But the objective in all cases has been to give a representation of natural language structures and arguments, so these points about pronouns are in the same category as those about predicates and their nominalisations, and sentences and their nominal-isations. It needs more than such a device as a lambda term, however, to adequately formalise reflexive pronouns. For pronouns are context-sensitive elements, and so a whole context-sensitive formal language is required to sym-bolise them. As before, there is no way to represent reflexive pronouns as such in a completely context-free language. Certainly a pronoun *with its antecedent* often can be represented in a context free way — thus 'a is not a member of itself' is the same as 'a is not a member of a'. But without that antecedent, the relevant context is unspecified, and the pronoun in the predicate 'is not a member of itself' is revealed to be a limited variable, i.e. the referent of the pronoun is seen to be functional upon the subject supplied.

II.E

Frege's handling of pronouns was therefore at fault, but also, when assimilating concepts to functions he was misled by a presumed affinity between identities and equivalences, and this had much larger consequences. Thus it was his move from 'x−y' to 'x>y', in Carnap's passage above — and, of course, similar moves in other passages — that led him astray. Certainly 'x−y' is a mathematical function of x and y, but 'x>y' is not. The former has two *arguments*, the latter two *subjects*, i.e. things the whole is saying something about. As a result the latter is a propositional, or logical function, of the form

$$\lambda z\lambda t(z>t)(y)(x),$$

and identifying the appropriate variables in it merely turns this into another propositional function with two subjects,

$$\lambda z\lambda t(z>t)(x)(x),$$

not a function of one variable

$$\lambda z(z>z)(x),$$

as with 'x-x'.

Returning to the original case of set-abstraction, we see that the predicate 'is not a member of itself' does not collect its subjects into a set, because there is a variable, 'itself', in this predicate, and so those things that are not members of themselves do not thereby have a common property — not even the common property of being members of the same set. Nevertheless, each one, paired with itself, is a member of a set of pairs. The Set Abstraction Axiom, in the case of elementary sentences, viz

$$(R^n)(\exists S)(x_1)(x_2) (x_n)(R^n x_1 x_2\ x_n \equiv <x_1, x_2,\ x_n> \in S),$$

thus requires that none of the variables are repeated in the general expression for the relation. Of course that still allows other equivalences to hold contingently in some cases, and even the given one when some variables are repeated — in the latter case they simply must be repeated, as well, in the ordered set on the right. Clearly similar revisions are necessary not only with n-ary relations, but also with second and higher-order ones, and comparable alterations must be made to Lambda Calculus abstracts.

LOGIC AND GRAMMAR

III.A

I have written a number of articles recently that have a rather remarkable character. They all point out trivial grammatical facts that, at great cost, have not been respected in twentieth century Logic. A major continuous strand in my previous work, with this same character, I will first summarise, to locate the kind of fact that is involved. But then I shall present an overview of the more recent, and more varied points I have made, which demonstrate the far larger extent of basic grammar that has been overlooked or suppressed. I end with some remarks about how this phenomenon can have arisen principally through logicians not being attentive enough to their own language, and occupying themselves, instead, with often quite imaginary languages.

III.B

As is explained more fully in Slater 2006c and 2009b (see also for instance 2005b) the main theme in my research over the past twenty years has been the linguistic reading of epsilon terms, and that shows the general line of my interests. Originally, I was very much taken with the work of Geach, on some problems in anaphoric reference. There are extensive structures in natural language such as linked pairs of sentences like 'Celia believed there was a woman in the room. But it was a man' which clearly cannot be handled using the logic coming down from Frege. Other notable authors working on this problem included Evans, who attempted no formalisation, and, for instance, Saarinen, who tried to work up a quite new one. One hindrance to getting the matter straight is illustrated in the case just given. For the pronoun 'it' there involves cross-reference into an intensional construction, which is not possible on Frege's views about indirect reference in such constructions. A better account, in this respect, was provided by Russell's Theory of Descriptions, although the possibility of 'quantifying in' with his 'primary sense' forms was found difficult to understand by Quine.

The Theory of Descriptions, in general, is an integral part of the problem with anaphora, as appeared in the later work of Neale in the area. But it was a variant to Russell's theory which attracted me, and which eventually led me to the solutions I subsequently proposed to a number of problems in this area. What had caught my eye, even as far back as 1963, was a feature of language that was highlighted more famously by Donnellan in 1966, and which I found in 1982 that Hilbert's epsilon terms could be used to express. Why does the epsilon account of definite descriptions work? It is largely because of a trivial point about grammar. If we put Russell's original case in anaphoric form, namely 'There is a single king of France, and he is bald', the 'he', it is easy to see, is going proxy for 'The king of France', and so, firstly, 'The king of France is bald' is not the whole remark, but merely its second conjunct. But then, secondly, the referent of its subject term is the object alluded to in the other conjunct, which means, thirdly, that that object exists not just in a world where the antecedent is true, since the cross-reference is secured independently of that. Another way of seeing this last, and major point is to consider the case where the second conjunct is, say, 'but he might not have been sole king of France'. That shows that the object referred to as 'The king of France' does not necessarily have the character attributed to it in that description. By incorporating the antecedent conjunct into his account of 'The king of France is bald' Russell came close to confusing reference with description, and objects with facts, and going with this he certainly formed the impression that this proposition could only be about an individual in a world where there was just one king of France. Later 'pre-suppositionists', while opposing Russell, held even more firmly to this opinion. But all this was despite it being a known logical truth, in Russell's logic, that $(\exists x)(x=a)$, for any individual a, which entails that such an individual has necessary existence. What was missing in this logic were Russell's 'logically proper names', which he could only describe informally; and it is such names that Hilbert's epsilon terms formalise. In particular they enable one to replace Russell's quantificational form

$(\exists x)[(y)(Ky \equiv y=x) \, \& \, Bx]$

with an equivalent one about an individual,

$(y)(Ky \equiv y=k) \, \& \, Bk,$

with $k=\epsilon x(y)(Ky \equiv y=x)$, and so see the above points formally. For it is 'Bk' which then represents the whole of 'the king of France is bald', and while it is contingently true that $(y)(Ky \equiv y=k)$, it is necessarily true that $(\exists x)(x=k)$.

These results enable us to acknowledge that we can talk about fictional objects just as well as factual ones, and that the same logic applies to each of them (Slater 2006b). Fictional objects simply do not live up to the description of them, in the actual world: The Gold Mountain, for instance, is an individual in all worlds, but only in some other world is it both gold and a mountain. In this world 'The Gold Mountain' is a misnomer, in other words, like Frege's example 'The Morning Star'. It is very ironic that Frege used this 'Millian' name so extensively in his illustrations, for its reference is not determined by its sense, as his theory required, because, of course, Venus is a planet. This possibly non-descriptive feature of referential phrases has many consequences. For instance, it gives an immediate resolution of Berry's Paradox, and related paradoxes, such as those discovered by Simmons (Slater 2005b).

The detail of the development of the above line of thought, however, is not what I want to concentrate on here, merely its general character. My concern was with the application of a certain well-established formal logic Hilbert's Epsilon Calculus to making more apparent grammatical structures and mechanisms in ordinary speech that had previously not drawn notice. Not all formal logicians have such a close interest in natural language, and not all students of natural language look at it formally. Hale and Wright, however, are two of the few contemporary philosopher-logicians with comparable interests, and I have written on matters close to their work in a number of recent papers. Thus in Slater 2003a, 2005a, 2005c and 2006d I have expanded my previous work to discuss the formalisation of mass and count terms, and shown its relevance to Frege's attempt to provide a logical foundation for Arithmetic. In natural language we distinguish 'it is gold' from 'it is a ring', for instance, but no such discrimination is present in Frege's concept script. Predicate Logic, therefore, covers both cases, but only in the latter case do numbers arise: there can be an amount of some stuff but not a number of it, and so nothing about numbers follows from Predicate Logic. In particular, $\neg(\exists x)Px$ is not equivalent to Nx:Px = 0, as Frege thought, since when 'P' is not count 'the number of Ps' is not defined. Hale and Wright are more aware than most of the count/mass distinction, but have payed no attention to this consequence of it. In the papers above I have expanded on this kind of point to provide a grammatical critique of set-theoretic analyses of number, and indeed of mathematical sets in their own right. Thus it is grammatical nonsense to say things like '{} = 0', since what is true instead is that Nx:(x $\in$ {}) = 0. And that point also shows that there are more things to be considered in Arithmetic than omega sequences, as Structuralists would conclude there are (lo and behold!) numbers, as well. In addition, a pair of apples, for instance, is not a further object besides the apples, as Set Theory has presumed. It is no more a further object than the half is in a half of a loaf of bread. 'Pairs' and 'halves' denote units of measure, and do not describe objects in their own right. Furthermore, groups of physical things are not 'sets' in the sense of Set Theory. For collective nouns like 'tribe', 'shoal', 'crowd', etc. describe mereological sums scattered objects that move around in space just like their members. Their members are thus physical parts of a further physical whole, and the relation between them is not 'set-membership' in the mathematical sense.

The general combination of logical and linguistic interests that I share with Hale and Wright I also shared with Prior, whose work on intensional constructions, for quite some time, was a major inspiration. But my recent work has also departed from Prior's line of thinking in one major respect. Prior was an operator theorist who tried to analyse locutions like 'a believed that p' as having the form 'Bap'. This follows traditional Modal Logic in which 'It is possible that p', for instance, is commonly symbolised 'Mp'. But Prior notably resisted the formalisation of such remarks in another way, as involving relations to, or properties of, propositions. Thus he argued against dividing 'a believes that p' into 'a believes' and 'that p', rather than 'a believes that' and 'p'. Davidson also argued for the latter kind of construal. Neither philosopher, therefore, was interested in the non-cleft forms of intensional and modal expressions, i.e. equivalent sentences like 'that p was believed by a', 'that p is possible'. It took Kneale, and Cocchiarella, in more recent times, to provide appropriate symbols for the 'that' in 'that p', i.e. for the grammatical element that converts a sentence into the kind of noun phrase which is embedded as the subordinate clause in intensional and modal constructions (Slater 2004b, 2005b). There is a resemblance between Kneale's and Cocchiarella's symbols, namely '§' and 'λ', and Montague's cap symbol, but Montague Grammar is too closely wedded to Fregean notions like sentences referring to truth-values, and identicals not being substitutable *salva veritate* in intensional contexts, to be entirely satisfactory.

The possibility of identicals being substitutable *salva veritate* in intensional constructions becomes evident once the proper theory of descriptions is employed, as I had shown in my previous work. If the ancients believed the Morning Star did not set in the evening, although it does so set, then they believed the Evening Star did not set in the evening. For the 'it' in the last sentence refers to the Morning Star, and so also to the Evening Star, since they are the same. Maybe, though, it wasn't exactly the Morning Star the ancients believed not to set in the evening, but instead something whose identity is not clear. In that case, of course, there is no possibility of substitution of identicals.

And do sentences refer to truth-values? That is a grammatical question that has been the focus of some of my more recent studies. Certainly the nominalisation of a sentence is a referring expression that is what 'nominalising a sentence' is for. But the nominalisation of a sentence is not that sentence itself, and so, in particular, we must be careful to distinguish sentences from 'that'-clauses. Remarkably, like Kneale and Cocchiarella subsequently, Frege did have originally a symbol for something like 'that', i.e. his 'horizontal', but he still could not see that, while 'that p' was referential, 'p' alone was not. Maybe that was because he also took predicates to be referential, since he took the reference of the whole sentence to be formed from the references of its parts. But it is the nominalisations of predicates, again, which are referential: it is 'being a horse' that refers to a concept, while 'is a horse' does not (Slater 2000c). Thinking otherwise got Frege into his famous problem about the concept of being a horse. For he thought concepts were unsaturated, and what were referred to by predicates in sentences. Certainly such predicates are unsaturated, needing subject terms to form a complete thought; but they are not referential, and what is referential in the area, namely their nominalisations, straightforwardly refer to certain saturated things, namely those abstract objects which are concepts. What Frege primarily lacked, in this area, therefore, was some symbol for the nominalisation of a predicate.

It is the lack of any symbol for sentence nominals, in the mainline tradition that followed him, which, maybe tempts some to still read the '⊃' in 'p ⊃ q' as 'materially implies', despite Quine's strictures. For 'implies' is a verbal relation requiring noun phrases on either side of it, as in the form 'that p materially implies that q', but '⊃' is a propositional connective, requiring, instead, used sentences adjacent to it, as in the case just given. The point demonstrates how elementary, yet how significant the piece of grammar is which is at the heart of this matter, at the same time as illustrating how possible is lack of attention to that piece of grammar. Maybe the distinction between connectives and verbs is often enunciated as a principle, but in practice, in such a common area as the reading of '⊃', this principle sometimes gets forgotten, or not applied.

Another, very similar case will show how such inattention to basic grammar was even what led to the most celebrated trouble with Fregean logic. Here, again, many would be likely to acknowledge in principle the relevant bit of grammar, although the full application of that principle has evidently not been pursued. I refer, now, to the difference between identity and equivalence. The expression for an identity, like 'a=b' is between two names, whereas the expression for an equivalence, like 'p ≡ q' is between two used sentences. Surely everyone knows that! But then why do they not object to Frege's identification of concepts with functions? Certainly one can say 'Pa ≡ T', and 'Rab ≡ F', where 'T' is some tautology and 'F' some contradiction, but these do not have the forms 'f(a)=T' and 'g(a, b)=F', where 'T' and 'F' abbreviate the referring phrases 'the true' and 'the false'. For equivalence is not identity (and sentences do not refer). I have dwelt on further matters in this area in Slater 2000b, 2002b, and 2004b. In fact the development of the proper grammar, together with a closer study of reflexive and other pronouns, then leads to a resolution of Russell's and related paradoxes.

What has not been noticed with such paradoxes is that there is a pronoun in the predicates 'is not a member of itself', 'does not apply to itself', 'yields a falsehood when appended to its own quotation'. Such variable predicates do not have a direct representation in a context independent language. Certainly one can represent in such a language the whole sentence 'x is not a member of itself', for instance, since this is just another way of writing 'x is not a member of x'. But that latter expression is a relational one of the form ' is R to ', not a predicative one of the form ' is P' with constant 'P'. In particular, the predicate in 'x is not a member of x' is not 'is not a member of a' with 'a' an item with a fixed reference, since the replacement for the pronoun, namely 'x', still varies with the subject. Nor is the predicate in 'x is not a member of x' the form of the whole sentence, namely 'y is not a member of y', since predicates are parts of, and not forms of sentences. If one wants to represent the whole as involving a predicate of x that predicate is not representable in a context insensitive language. Only a relational analysis is possible in such a language. And what has also been little noticed is that there is no paradox if 'x is not a member of x' is analysed relationally.

One can locate Frege's crucial false move that led to the generation of Russell's Paradox on this basis. The false move originates in Frege's attachment to Mathematics. For it is well known that it was Russell's glaring contradiction that alerted Frege to the trouble with his system; but it is less well known that there is less trouble when 'x is not a member of x' is analysed relationally, i.e. as saying that <x,x> is a member of {<y,z> : y is not a member of z}. For substitution of that set abstract for 'x' does not produce a similar contradiction. What Russell's Paradox shows, in fact, is merely that not all relations between a thing and itself can be a matter of that thing falling under a concept, i.e. having a constant property: $\neg(R)(\exists P)(x)(Rxx \neg Px)$. But clearly, given a two-valued function $f(x,y)$, one can invariably obtain a function of one variable $f(x,x)=g(x)$. So it was Frege's analogy between functions and predicates, in 'Function and Concept', which led him astray. Predicates are not functions in the required way. First, as we have seen, sentences are not referential terms with the same reference as 'the true' or 'the false'. What one has are equivalences like 'Pa≡T' and 'Rab≡F' which are themselves equivalent to 'Pa' and '¬Rab' respectively, making 'Pa' and 'Rab' quite unlike mathematical functions, and 'T' and 'F' nothing like their values. Certainly we say such things as 'it is true that Pa' and 'it is false that Rab', but these expressions predicate certain properties of thoughts they might be represented as 'T<Pa>', and 'F<Rab>', with 'T' now 'is true', and 'F' now 'is false'. So here again there is no reference to truth values, since as we have seen, predicates are not referring expressions.

But Frege's grammatical inaccuracies about concepts and functions, and about predicates and referring phrases, get dramatically enlarged upon the introduction of reflexive expressions. For not all predicates express a constant concept. If each of A, B, and C shaves D they do the same thing shave D but if each shaves himself, or say, in a ring, shaves his neighbor on his left, then they only do the same kind of thing, i.e. what they do merely has a common functional expression: shave f(s) where s is the subject. There might be a contingent equivalence between 'x shaves himself' and 'x is either A or B or C', if all and only A, B, and C shave themselves. But 'x shaves himself' is not *logically equivalent* to any such set-membership expression, because it's predicate contains a variable, 'himself', allowing there to be 'paradoxical' cases where there is no set at all.

In other areas the incorporation of symbols like Frege's horizontal also has considerable significance. Indeed, it enables Tarski's theory of Truth to be corrected grammatically which has the immediate consequence that the Liar Paradox is then not a problem (Slater 2000b, 2001, 2002b, 2004b). Related to this similar points can be made with respect to the syntactic theory of provability, where the comparable puzzles lie in Gödel's Theorems. Here again we find that lack of attention to basic grammar is at the heart of the matter. Specifically, there have been difficulties representing propositional referring phrases of the form 'that p', and that has meant that what is to be proved in Mathematics has not been clear. Many essays on the Foundations of Mathematics have held that such a formula as '2 + 3 = 5' is true in the standard model of Arithmetic, whereas what is true there is that 2 + 3 = 5; and formal derivations of the formula '2 + 3 = 5' from other formulas do not suffice to prove that 2 + 3 = 5. Using ' p ' as an alternative to '<p>', confuses syntactic expressions with their immediate readings, and removing that confusion reduces the interest in Gödel's Theorems in several ways, principally by undermining the nominalistic philosophy that selected derivations of quoted formulas as the paradigm of a mathematical proof. Moreover, unlike with Gödel's derivability predicate, where the formula 'p' is not derivable from 'Bew "p" ', from the fact that it is provable that p it does follow that p, and that shows that the consistency of provability is immediate. These points are intimately related to Wittgenstein's views on the Foundations of Mathematics, I believe, and not just in connection with his neglect of Gödel's results.

Is it believable, though, when it has such enormous consequences, that inattention to such obvious and trivial pieces of grammar as the difference between ' "p" ' and '<p>', is rife, even among learned and careful men? Here are just two recent cases where there has been inattention to this instance of the otherwise well enunciated difference between use and mention. First there is Stephen Read (Read 2008 sec. 1), who gives Tarksi's T-scheme as:

(T) x is true if and only if p,

where what replaces 'x' is a name of a sentence whose translation into the metalanguage replaces 'p'. But Read then goes on to quote Horwich as affirming (T) as a truism, when saying "for any declarative sentence 'p' our language generates an equivalent sentence 'The proposition that p is true' ". Read adds further 'Tarski did not propose (T) as a definition of truth, though others, e.g. Horwich, have done so since. They all describe (T) as a truism'. But Read is not the only learned and careful man who confuses use with mention in this area. There is also John Burgess, who says (Burgess 2005 p 205): "first, truth and falsehood conform to a disquotation scheme, according to which it is true that p if and only if p, and false that p if and only if not p. Second, as an immediate consequence of this first feature plus the classical principle of the excluded middle, according to which we always have either p or not p, truth conforms to the principle of bivalence, according to which it is always either true that p or false that p." (N.B. 'it is true that p' does not involve quotation).

How can these kinds of thing have gone on, in such a wide range of places, over such a length of time, and with such a number of learned and careful people? Confusing use with mention, identity with equivalence, predicates with functions, predicates with their nominalisations, sentences with nominals, sentences with propositions, reference with description, pronouns with nouns, mereological sums with (mathematical) sets, numbers with sets, count terms with mass terms, amounts with numbers, units of measure with objects and propositions with functions, propositions with sets, as we shall see displays a major failure in the disposition of the logicians who have followed on from Frege. But how has this general phenomenon come about? One prime reason surely relates to the mathematisation of Logic in the same period. Not only was Frege primarily a mathematician, but there has developed since his time a branch of the subject now called 'Mathematical Logic', which has generated extensive results in Meta-theory as well as Set Theory. We have seen how it was Frege's orientation towards Mathematics that led him to confuse predicates with functions, but there is also a well-known general disjunction between mathematical talent and verbal ability, which easily can lead, for instance, to the acceptance of malformed identities like '{}=0' without a moment's question. Intelligence is clearly not linked to linguistic competence; and the downside of that is that very brainy people can easily have blind spots in areas like grammar.

The belief that professionalism in Logic requires advanced mathematical capacity is to be found in several 'theories of propositions' I have not discussed before in print. An inspection of these 'theories' will serve very well to demonstrate why the required capacity is instead the ability to speak in natural language. One group of these theorists, for instance, says that propositions are functions from possible worlds to truth-values. That might sound impressively professional until one reflects that a possible world cannot be given independently of the propositions that are true in it. Maybe there are other worlds in which the proposition that p is true, but that does not mean one can first give those worlds in some way, and then work out, via some independently presentable function, whether the proposition that p is true in them. Furthermore, if the proposition were the function then that function of worlds would have to be also what was true in the appropriate worlds. For a given proposition one might define a function F such that $F(i)=1$ if $V(p, i)=1$, and $F(i)=0$ if $V(p,$

i)=0, but there is no way, then, that the proposition is F, since 'V(F, i)=1' does not make sense.

In another tradition it is said that propositions are sets: a proposition is said to be the set of worlds in which it is true. This kind of definition might be appealing to somebody who wants to believe the Ontology of the world consists solely of sets, but it suffers from the same problem as before. For the set of worlds in which something is true is not itself something with a truth value which varies from one world to the next: 'V({j : V(p, j)=1}, i)=1' does not make sense. Indeed there is another major problem in this case, because 'the proposition that p is the set of worlds in which the proposition that p is true' is circular, and so defines nothing.

A third 'theory' of propositions is even well known nonsense. For other set-theorists want to say that an elementary proposition is an ordered set with n objects and an n-place relation as members: $<a_1, a_2, a_n, R^n >$. This expression evidently has the advantage, for some, that it is in the currently most respected formal language; but it has the overriding disadvantage, should anyone care to notice it, that nothing in that language is in the right category to do the job. Not only, in addition to sets are there numbers; also, in addition to them, there are propositions! Why does one need something in a different category? Because no set can be stated; in particular, one cannot state $<a_1, a_2, a_n, R^n >$. One can state, for instance, that Dobbin is a horse, but one cannot state: <Dobbin, being a horse>. The problem with the latter entity has even been given a name, it is so celebrated: it is 'The Problem of the Unity of the Proposition'. But the now century-old question of how, using a list of names, one can state something has not impeded the repeated enunciation of this third 'theory'.

Indeed, from within none of the above mathematical traditions has the reality check of a comparison with expressions of the form 'that p' been attempted. Nowhere in them is such an entity as that Dobbin is a horse discussed. Such an entity is therefore not recognised as a possible fact, i.e. as just what the various 'theories' are supposed to be about. And so we see why I have put the word 'theory' in scare quotes in all of the above. For once what the 'theories' are supposed to be about is recognised, it is immediately evident that no 'theory' of those things was what was needed, merely a non-mathematical, natural language in which to identify them.

That leads us to the most important reason, I believe, why there has been such a massive blindness in the area. For the main reason, surely, is that the points I have drawn attention to involve certain aspects of natural language, and a division has arisen, over the last 100 years, between a large portion of 'Logic' and the study of that. The interest in Meta-theory, and particularly Tarskian Semantics, is a major cause of this division, since it has distanced the language studied from the language used to study it, and so has led not only to inattention to that used language, but also encouraged entirely idle attention to quite unusable languages. Isn't the application of such a form as ' "p" is true' required for us to understand 'the object language' from our 'meta-linguistic' perspective? Not if the 'object language' is our Mother Tongue, and the understanding of it is instead provided though use. Davidson used Tarski's

T-scheme as a basis for a theory of interpretation, but when assessing as true something said in one's own language the meaning is already given. Hence the operative form of the judgement is 'it is true that p' or 'that p is true'. The problem has been formulating reference to the meaning of 'p'. For it is not anything of the syntactic form ' "p" ' that is being judged, but its immediate semantic reading, 'that p'.

What else, in particular, could be necessarily true? Certainly not some syntactic form. When judging some argument, like Disjunctive Syllogism, to be valid, one is not assessing bare formulas like '[¬p & (p v q)] ⊃ q' for some quality. Instead one is asking about a certain interpretation of them, and it is that interpretation of them which is necessarily true. The use-mention confusion involved in thinking otherwise has been a major feature of recent 'logic'. Many logicians in the twentieth century were very much concerned with what they saw as alternatives to 'classical logic', starting with the Intuitionists who said they doubted the Law of the Excluded Middle. Certainly the formula 'p v ¬p' is not a provable formula in their system, but that casts no doubt onto the classical law. For 'The Law of the Excluded Middle' refers not to the formula, but to its classical interpretation. One might not know the interpretation of some symbol, and for that, or other reasons, have doubts about whether the formula expresses a necessary truth, but that is not a matter of doubting the necessary truth itself. The problem in getting this clear has been, again, the lack, within formal logic, of appropriate denoting phrases for what it is that is necessary, although just such are the truths of Logic.

BIBLIOGRAPHY

[1] B. Armour-Garb, and J.A. Woodbridge, *Dialetheism, Semantic Pathology and the Open Pair*, Australasian Journal of Philosophy **84** (2006), 395–416.

[2] R. Bambrough, *Moral Skepticism and Moral Knowledge*, Routledge and Kegan Paul, London, 1979.

[3] G. Baker, and P. Hacker, *Ostensive Definition and its Ramifications*, Wittgenstein: Understanding and Meaning, Blackwell, Oxford, 1980, Ch V.

[4] P. Bernays, *Axiomatic Set Theory*, North-Holland, Amsterdam,1968.

[5] R. Bertolet, *The Semantic Significance of Donnellan's Distinction*, Philosophical Studies **37** (1980), 281–8.

[6] M. Black, *Making Something Happen*, Determinism and Freedom (S. Hook (ed.)), Collier Books, New York, 1958.

[7] J. Burgess, *Review of 'The Taming of the True' by N. Tennant*, Philosophia Mathematica **13** (2005), 202–215.

[8] R. Carnap, *On the Use of Hilbert's Epsilon Operator in Scientific Theories*, Essays on the Foundations of Mathematics (Y. Bar-Hillel etal. (eds)), The Magnes Press, Jerusalem 1961, pp156-164.

[9] L. Cheung, *The Tractarian Operation N and Expressive Completeness*, Synthese **123** (2000), 247–162.

[10] N. Cocchiarella, *Logical Investigations of Predication Theory and the Problem of Universals*, Bibliopolis, Naples, 1986.

[11] N. Cocchiarella, *Logical Studies in Early Analytic Philosophy*, Ohio State University Press, Columbus Ohio, 1987.

[12] I. Copi, *Symbolic Logic*, Macmillan, New York, 1973.

[13] D. Davidson, *On Saying That*, Synthese **19** (1968), 130–146.

[14] D. Davidson, *Inquiries into Truth and Interpretation*, Clarendon, Oxford, 1984.

[15] M. Devitt, *Singular Terms*, Journal of Philosophy **71** (1974), 183–205.

[16] K. Donnellan, *Reference and Definite Descriptions*, Philosophical Review **75** (1966), 281–304.

[17] M. Dummett, *Platonism*, Truth and Other Enigmas, Duckworth, London, 1978.

[18] U. Egli, and K. von Heusinger, *The Epsilon Operator and E-type pronouns*, Lexical Knowledge in the Organization of Language (U. Egli, et al, (eds.)) Benjamins, Amsterdam, 1995, pp. 1–24.

[19] G. Evans, *Pronouns, Quantifiers and Relative Clauses*, Canadian Journal of Philosophy **7** (1977), 567–636.

[20] S. Feferman, *Towards Useful Type-free Theories 1*, Recent Essays on Truth and the Liar Paradox (R.L. Martin (ed.)), O.U.P., Oxford, 1984.

[21] A. Flew, *Essays in Conceptual Analysis*, Greenwood, Westport, 1966.

[22] R.J. Fogelin, *Wittgenstein*, Routledge and Kegan Paul, London, 1976.

[23] R.J. Fogelin, *Wittgenstein on Identity*, Synthese **56** (1983), 141–154.

[24] G. Frege, *Translations from the Philosophical Writings of Gottlob Frege* (P.T. Geach and M. Black (eds)), Blackwell, Oxford, 1952.

[25] G. Frege, *The Foundations of Arithmetic* (J.L. Austin, (ed.)), Blackwell, Oxford, 1968.

[26] G. Frege, *Conceptual Notation and Related Articles* (T.W. Bynum (ed.)), Clarendon, Oxford, 1972.

[27] G. Frege, *Frege's Lectures on Logic* (E.H. Reck and S. Awodey (eds)), Open Court, Chicago, 2004.

[28] P. Frascolla, *Wittgenstein's Philosophy of Mathematics*, Routledge, London, 1994

[29] P.T. Geach, and M. Black, (eds), *Translations from the Philosophical Writings of Gottlob Frege*, Blackwell, Oxford, 1952.

[30] P.T. Geach, *Reference and Generality*, Cornell, Ithaca, 1962.

[31] P.T. Geach, *Intentional Identity*, Journal of Philosophy **64** (1967), 627–32.

[32] P.T. Geach, *Logic Matters*, Blackwell, Oxford, 1972.

[33] L. Goddard, and R. Routley, *The Logic of Significance and Context*, Scottish Academic Press, Aberdeen, 1973.

[34] L. Goldstein, *'This Statement is not True' is not True*, Analysis **52** (1992), 1–5.

[35] L. Goldstein, *Fibonacci, Yablo and the Cassationist Approach to Paradox*, Mind **115** (2006), 867–890.

[36] R.L. Goodstein, *On the Formalisation of Indirect Discourse*, Journal of Symbolic Logic **23** (1958), 417–419.

[37] S. Haack, *Philosophy of Logics*, C.U.P., Cambridge, 1978.

[38] B. Hale, and C. Wright, *The Reason's Proper Study*, Clarendon, Oxford, 2001.

[39] D. Hilbert, and P. Bernays, *Grundlagen der Mathematik*, Springer, Berlin, 1970.

[40] J. Hintikka, *Identity, Variables, and Impredicative Definitions*, Journal of Symbolic Logic **21** (1956), 225–254.

[41] J. Hintikka, *Knowledge and Belief*, Cornell University Press, Ithaca, 1962.

[42] P. Horwich, *Truth, 2nd Ed.*, Clarendon, Oxford, 1998.

[43] G.E. Hughes, and M.J. Cresswell, *An Introduction to Modal Logic*, Methuen, London, 1968.

[44] D. Kalish, and R. Montague, *Logic; Techniques of Formal Reasoning*, Harcourt, Brace and World, New York, 1964.

[45] D. Kaplan, *Demonstratives*, Themes from Kaplan (J. Almog, J. Perry and H. Wettstein (eds)), O.U.P., Oxford, 1989.

[46] W. Kneale, *Propositions and Truth in Natural languages*, Mind **81** (1972), 225–243.

[47] W. Kneale, and M. Kneale, *The Development of Logic* Clarendon, Oxford, 1962.

[48] G.T. Kneebone, *Mathematical Logic and the Foundations of Mathematics*, Van Nostrand, Dordrecht, 1963.

[49] S. Kripke, *Naming and Necessity*, Semantics of Natural Language (D. Davidson and G. Harman (eds)), Reidel, Dordrecht, 1972.

[50] S. Kripke, *Outline of a Theory of Truth*, Journal of Philosophy **72** (1975), 690–716.

[51] A. Koslow, *A Structuralist Theory of Logic*, C.U.P., Cambridge, 1972.

[52] A.C. Leisenring, *Mathematical Logic and Hilbert's Epsilon Symbol*, Macdonald, London, 1969.

[53] H. Leitgeb, *Formal and Informal Provability*, New Waves in Philosophy of Mathematics (O. Bueno and O. Linnebo (eds.)), Palgrave Macmillan, New York, 2009.

[54] E.J. Lemmon, *Sentences, Statements and Propositions*, British Analytical Philosophy (B. Williams and A. Montefiore (eds)), Routledge and Kegan Paul, London, 1966.

[55] D.K. Lewis, *On the Plurality of Worlds*, Blackwell, Oxford, 1986.

[56] B. McGuinness, (ed.), *Gottlob Frege: Philosophical and Mathematical Correspondence*, University of Chicago Press, Chicago, 1980.

[57] W.P.M. Meyer Viol, *Instantial Logic*, ILLC, Amsterdam, 1995.

[58] R. Montague, *Formal Philosophy* (R. H. Thomason (ed.)), Yale University Press. New Haven, 1974.

[59] S. Neale, *Descriptions*, M.I.T. Press, Cambridge MA, 1990.

[60] S. Neale, *The Philosophical Significance of Gödel's Slingshot*, Mind **104** (1995), 761–823.

[61] C. Parsons, *Mathematical Intuition*, Proceedings of the Aristotelian Society **80** (1979-80).

[62] C. Parsons, *Intuition in Constructive Mathematics*, Language, Mind and Logic (J. Butterfield (ed.)), C.U.P., Cambridge, 1986.

[63] J. Passmore, *Philosophical Reasoning*, Duckworth, London, 1961.

[64] R. Penrose, *The Emperor's New Mind*, Vintage, London, 1990.

[65] G.G. Priest, *Indefinite Descriptions*, Logique et Analyse **22** (1979a), 5–21.

[66] G.G. Priest, *The Logic of Paradox*, Journal of Philosophical Logic **8** (1979b), 219–241.

[67] G.G. Priest, *Wittgenstein's Remarks on Gödel's Theorem*, Wittgenstein's Lasting Significance (M. Köbel and B. Weiss (eds)), Routledge, London, 2004.

[68] G.G. Priest, *Towards Non-Being: The Logic and Metaphysics of Intentionality*, O.U.P., Oxford, 2005.

[69] G.G. Priest, *Doubt Truth to be a Liar*, O.U.P., Oxford, 2006.

[70] G.G. Priest, *Against against non-being*, The Review of Symbolic Logic **4** (2011), 237–253.

[71] A.N. Prior, *Formal Logic*, O.U.P, Oxford, 1962.

[72] A.N. Prior, *Objects of Thought*, O.U.P., Oxford, 1971.

[73] S. Psillos, *Rudolf Carnap's Theoretical Concepts in Science*, Studies in History and Philosophy of Science **32** (2000), 151-172.

[74] W.C. Purdy, *A Variable-free Logic for Anaphora*, Patrick Suppes Scientific Philosopher Vol 3 (P. Humphreys (ed.)), Kluwer, Dordrecht, 1994, 34–57.

[75] W.V.O. Quine, *Methods of Logic, Rev. Ed.*, Holt, Rinehart and Winston, New York, 1959.

[76] W.V.O. Quine, *Quantifiers and Propositional Attitudes*, Reference and Modality (L. Linsky (ed.)), O.U.P., Oxford, 1971.

[77] C. Radford, *How can we be moved by the fate of Anna Karenina?*, Proceedings of the Aristotelian Society **49** (1976), 67–93.

[78] F.P. Ramsey, *The Foundations of Mathematics*, Foundations (D. H. Mellor (ed.)), Routledge and Kegan Paul, London, 1978.

[79] S. Read, *The Liar Paradox from John Buridan back to Thomas Bradwardine*, Vivarium **40** (2002), 189–218.

[80] S. Read, *The Truth Schema and the Liar*, Unity, Truth and the Liar (S. Rahman, T. Tulenheimo and E. Genot (eds)), Springer, Berlin, 2008a.

[81] S. Read, *Further Thoughts on Tarski's T-scheme and the Liar*, Unity, Truth and the Liar (S. Rahman, T. Tulenheimo and E. Genot (eds)), Springer, Berlin, 2008b.

[82] S. Read, *Field's paradox and Its Medieval Solution*, History and Philosophy of Logic **31** (2010), 161–176.

[83] C. Reid, *Hilbert*, Springer-Verlag, Berlin, 1970.

[84] R. Routley, *Choice and Descriptions in Enriched Intensional Languages 1, 2, 3*, Problems in Logic and Ontology (E. Morscher, J. Czermak, and P. Weingartner (eds.)), Akademische Druck- und Velagsanstalt, Graz, 1977.

[85] R. Routley, R. Meyer, and L. Goddard, *Choice and Descriptions in Enriched Intensional Languages 1*, Journal of Philosophical Logic **3** (1974), 291-316.

[86] E. Saarinen, *Intentional Identity Interpreted*, Game Theoretical Semantics (E. Saarinen (ed.)), Reidel, Dordrecht, 1978, pp. 245–327.

[87] R. Shiner, *Wittgenstein and the Foundations of Knowledge*, Proceedings of the Aristotelian Society **LXXVIII** (1977-8), 103-124.

[88] B.H. Slater, *Talking about Something*, Analysis **93** (1963), 49–53.

[89] B.H. Slater, *Is 'heterological' heterological?*, Mind **327** (1973), 439–40.

[90] B.H. Slater, *Sensible Self-Containment*, Philosophical Quarterly **34** (1984), 163–4.

[91] B.H. Slater, *E-type Pronouns and Epsilon Terms*, Canadian Journal of Philosophy **16** (1986), 27-38.

[92] B.H. Slater, *Fictions*, British Journal of Aesthetics **27** (1987), 145–55.

[93] B.H. Slater, *Hilbertian Reference*, Nous **22** (1988), 283–97.

[94] B.H. Slater, *Descriptive Opacity*, Philosophical Studies **66** (1992a), 167–81.

[95] B.H. Slater, *Routley's Formulation of Transparency*, History and Philosophy of Logic **13** (1992b), 215-24.

[96] B.H. Slater, *The Logical Paradoxes*, The Internet Encyclopedia of Philosophy (2000a), (http://www.utm.edu/research/iep/p/par-log.htm).

[97] B.H. Slater, *The Uniform Solution*, LOGICA Yearbook 1999, Czech Academy of Sciences, Prague, 2000b.

[98] B.H. Slater, *Concept and Object in Frege*, Minerva **4** (2000c) (http://www.ul.ie/ philos/vol4/index.html)

[99] B.H. Slater, *Prior's Analytic Revised*, Analysis **61** (2001a), 86–90.

[100] B.H. Slater, *Epsilon Calculi*, The Internet Encyclopedia of Philosophy (2001b) (http://www.utm.edu/research/iep/ep-calc.htm)

[101] B.H. Slater, *Syntactic Liars*, Analysis **62** (2002a), 107–9.

[102] B.H. Slater, *Namely-Riders: an Update*, Electronic Journal of Analytic Philosophy (2002b) (http://ejap.louisiana.edu/EJAP/2002/Slater.html).

[103] B.H. Slater, *Choice and Logic*, Logical Studies **9** (2002c) (http://www.logic.ru/LogStud/09/No9-06.html)

[104] B.H. Slater, *Aggregate Theory versus Set Theory*, Erkenntnis **59** (2003a), 189–202

[105] B.H. Slater, *Tarski's Hidden Assumption*, Ratio **XVII** (2003b), 84–89.

[106] B.H. Slater, *Ramseying Liars*, Logic and Logical Philosophy **13** (2004a), 57–70.

[107] B.H. Slater, *A Poor Concept Script*, Australasian Journal of Logic (2004b) (http://www.philosophy. unimelb.edu.au/ajl/).

[108] B.H. Slater, *Ramsey's Tests*, Synthese **141** (2004c), 431–444.

[109] B.H. Slater, *Logic and Arithmetic*, LOGICA Yearbook 2004, Czech Institute of Philosophy, Prague, 2005a.

[110] B.H. Slater, *Choice and Logic*, Journal of Philosophical Logic **43** (2005b), 207–216.

[111] B.H. Slater, *Aggregate Theory versus Set Theory*, Proceedings of the PILM Conference Nancy 2002, Philosophia Scientiae, Nancy, 2005.

[112] B.H. Slater, *Frege's Hidden Assumption*, Critica **38** (2006a), 27–37.

[113] B.H. Slater, *The Epsilon Logic of Fictions*, Mistakes of Reason: Essays in Honour of John Woods (K. Peacock and A. Irvine (eds)), University of Toronto Press, Toronto, 2006b, pp. 33–48.

[114] B.H. Slater, *Grammar and Sets*, Australasian Journal of Philosophy **84** (2006c), 59–73.

[115] B.H. Slater, *Logic and Arithmetic*, Perspectives in Logicism (P. Joray (ed.)), Travaux de Logique, Neuchatel, 2006d.

[116] B.H. Slater, *Epsilon Calculi*, Logic Journal of the IGPL **14.4** (2006e), (http://jigpal.oxfordjournals.org/cgi/content/full/14/4/535).

[117] B.H. Slater, *Logic and Grammar*, Ratio **XX** (2007a), 206–218.

[118] B.H. Slater, *The Central Error in the Tractatus*, Wittgenstein Jahrbuch 2003/2006, Peter Lang, Bern, 2007b, 57-66.

[119] B.H. Slater, *Completing Russell's Logic*, Russell **27** (2007c), 78–92.

[120] B.H. Slater, *Harmonising Natural Deduction*, Synthese **163** (2008a), 187-198.

[121] B.H. Slater, *Out of the Liar Tangle*, Unity, Truth and the Liar (S. Rahman, T. Tulenheimo and E. Genot (eds)), Springer, Berlin, 2008b.

[122] B.H. Slater, *A Poor Concept Script*, Dimensions of Logical Concepts (J-Y.Beziau and A.Costa-Leite (eds.)), Colecao CLE, Campinas, 2009a.

[123] B.H. Slater, *Hilbert's Epsilon Calculus and its Successors*, Handbook of the History of Logic Vol 5 (D. Gabbay and J. Woods (eds)), Elsever Science, Burlington MA, 2009b, 385–448.

[124] B.H. Slater, *What Priest (amongst many others) has been missing*, Ratio **XXIII** (2010a), 184–198.

[125] B.H. Slater, *Ontological Discriminations*, LOGICA Yearbook 2009, Czech Institute of Philosophy, Prague, 2010b.

[126] B.H. Slater, *Quine's other way out*, The Reasoner **4.3** (2010c), 37–8.

[127] B.H. Slater, *Back to Aristotle!*, Logic and Logical Philosophy (2011a).

[128] B.H. Slater, *Translatable Self-Reference*, Australasian Journal of Logic (2011b) (http://www.philosophy.unimelb.edu.au/ajl/).

[129] B.H. Slater, *Quine's other way out*, LOGICA Yearbook 2010, Czech Institute of Philosophy, Prague, 2011c.

[130] B.H. Slater, *Natural Language Consistency*, Logique et Analyse (2011d).

[131] R. Stalnaker, *Ways a World might be: Metaphysical and Anti-metaphysical Essays*, Clarendon, Oxford, 2003.

[132] R. Sorensen, *A Definite No-No*, Liars and Heaps (J. C. Beall (ed.)), O.U.P, Oxford, 2003.

[133] A. Tarski, *Logic, Semantics and Metamathematics* (J. H. Woodger (ed.)), Clarendon, Oxford, 1956.

[134] R Thomason, and R. C. Stalnaker, *Modality and Reference*, Nous **2** (1968), 359–372.

[135] M. Thompson, *On Aristotle's Square of Opposition*, Philosophical Review **62** (1953), 251–265.

[136] T. Williamson, *Self-Knowledge and Embedded Operators*, Analysis **56** (1990), 202-209.

[137] L. Wittgenstein, *Tractatus Logico-Philosophicus* (D. F. Pears and B. F. McGuinness (eds)), Routledge and Kegan Paul, London, 1961.

[138] L. Wittgenstein, *Remarks on the Foundations of Mathematics, 3rd edn.* (G. H. von Wright, R. Rhees and G. E. M. Anscombe (eds)), Blackwell, Oxford, 1978.

[139] C. Wright, *Frege's Conception of Numbers as Objects*, Aberdeen University Press, Aberdeen, 1983.

[140] S. Yablo, *Paradox without Self-Reference*, Analysis **53** (1993), 252–3.

INDEX

Lightning Source UK Ltd.
Milton Keynes UK
UKOW03f2114071013

218638UK00010B/1077/P